Near Field Communications Handbook

INTERNET and COMMUNICATIONS

This new book series presents the latest research and technological developments in the field of Internet and multimedia systems and applications. We remain committed to publishing high-quality reference and technical books written by experts in the field.

If you are interested in writing, editing, or contributing to a volume in this series, or if you have suggestions for needed books, please contact Dr. Borko Furht at the following address:

Borko Furht, Ph.D.
Department Chairman and Professor
Computer Science and Engineering
Florida Atlantic University
777 Glades Road
Boca Raton, FL 33431 U.S.A.

E-mail: borko@cse.fau.edu

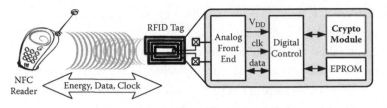

Figure 2.1   The basic scheme of a passive RFID system.

portable NFC device), which may have an online connection to a back-end database or may work as a stand-alone, generates an electromagnetic field. The energy is transmitted from the reader over the air interface to the tag. This passive RFID tag, which has no internal power source such as a battery, draws the required energy from the electromagnetic field. The bidirectional communication works in a half-duplex manner. The reader uses some kind of amplitude-shift keying (ASK) modulation technique for sending data to the tag. The tag-to-reader communication works by modulating the continuous-wave signal of the reader by the tag.

The tag architecture depicted in Figure 2.1 shows the main parts of the integrated circuit (IC) that is attached and connected to the antenna. The analog front-end is responsible for the power supply of the tag, the bidirectional communication using modulation and demodulation, as well as for clock signal generation. In HF systems, the clock signal is recovered from the carrier frequency of the field. The digital control unit defines the functionality of the tag. It decodes the data from the demodulator according to the used RFID air interface standard and answers by controlling the modulator circuit. It performs the anticollision sequence for bulk detection of tags and provides access to the tag's nonvolatile memory. The nonvolatile memory is mostly realized as EEPROM and contains data such as the unique identifier (UID), which has to be retained when the tag is not powered by a reader. The presented scheme of an RFID tag has an additional module that supports cryptographic functionality. This is not typical for existing RFID tags, which are solely for the purpose of identification, but it emphasizes the intended goal of our work to enhance the functionality of passive RFID tags.

Applications of NFC/RFID Technology

Originally, RFID technology was invented to replace barcodes. The main advantage over barcodes is that there is no need for a clear line

Near Field Communications Handbook

Edited by
Syed A. Ahson and Mohammad Ilyas

CRC Press
Taylor & Francis Group
Boca Raton London New York

CRC Press is an imprint of the
Taylor & Francis Group, an **informa** business

CRC Press
Taylor & Francis Group
6000 Broken Sound Parkway NW, Suite 300
Boca Raton, FL 33487-2742

First issued in paperback 2019

ISBN-13: 978-1-4200-8814-4 (hbk)
ISBN-13: 978-0-367-38236-0 (pbk)

Visit the Taylor & Francis Web site at
http://www.taylorandfrancis.com

and the CRC Press Web site at
http://www.crcpress.com

Contents

Preface

Near Field Communication (NFC) is a short-range, high-frequency wireless communication technology that has emerged from the convergence of contactless identification such as RFID and networking technologies such as Bluetooth and Wi-Fi. An NFC communication can be initialized by as simple a means as "touching" or "tapping" one NFC device to another. Two NFC-compatible devices communicate (exchange data) over a distance of a few centimeters. Such a short transmission range makes eavesdropping inherently hard, and thus helps achieve better communication security. NFC enables direct interaction between mobile devices such as phones or cameras as well as allowing access to RFID-tagged items and contactless smart cards. Massive growth in the industry started in the late 1990s with the possibility of fabricating low-cost RFID tags in high volume. NFC allows access to those tags not only from dedicated reader devices, but on the basis of devices we carry with us every day (e.g., a mobile phone or camera); it will represent therefore the major interface between our daily life and tagged objects. The basic principle of RFID technology is that a device called a "reader" or "interrogator"—in our case a mobile phone with NFC interface—provides an RF field. It transmits data to the tag, which is also known as the transponder, by modulating the field. The tag is an integrated circuit that is attached to an antenna. It receives requests and transmits data to the reader.

Over the last 15 years, the growing availability of wireless communication technologies, as well as the miniaturization of electronic components inside consumer devices, has enabled the possibility of creating a so-called ubiquitous computing environment, as foreseen by Mark Weiser at the beginning of the 1990s. In a Weiser article titled "The Computer for the 21st Century," the researcher from XEROX PARC hypothesized a world of interconnected objects by means of information and communication technologies. NFC technology enables local communication between entities in our everyday environment. These features facilitate creating a network of interconnected objects known as the "Internet of Things." NFC focuses on user-initiated communication, due to the short reading distance a user has to bring a reader and a tag near each other. As NFC technology becomes common, various electrical devices can be equipped with NFC readers. These readers can then be used to read and write RFID tags and even to communicate with other devices equipped with similar readers.

In the meantime, another important process related to ICT has appeared. It deals with the convergence of different communication technologies inside one single device, the mobile phone, nowadays spread among 61% of the worldwide population, according to the International Telecommunication Union (ITU). Suitable to be used with a variety of devices, NFC technology is mainly aimed at mobile phones. The computational power of modern mobile phones and NFC wireless connectivity technology has become the enabler of various mobile-based applications, such as electronic ticketing for public transportations, mobile payment, and bootstrapping Wi-Fi or Bluetooth-pairing processes. The convergence of a number of communication interfaces (Bluetooth, Wi-Fi, W-CDMA, and NFC, just to mention some) inside the mobile phone, makes this device the most qualified for accessing different types of services into an interconnected environment such as the one predicted by Mark Weiser. The centrality of the mobile phone inside everyone's life is one of the reasons why NFC technology has caught the attention of different industries and research institutions. In fact, NFC can be seen as the integration of Radio Frequency Identification (RFID) technology inside the mobile phone, potentially allowing more than a half of the worldwide population to interact with smart environments via RFID technology.

NFC will enable a range of applications and services: proximity services such as getting information by touching smart posters or sharing information between phones or any NFC devices, being the boarding pass in an airport, functioning as a remote controller or a key, advertising services, location-based services, and so forth. To achieve these services and many more we may not yet have even thought of, it should be highlighted that PCs, TVs, other computing devices and noncomputing devices should become NFC enabled. We believe in the long-term potential of NFC to make a significant difference in the quality of life and comfort for many people and for our environment. This technology can be woven itself into our environment according to Weiser's vision of Ubiquitous Computing, and it can be the key technology for the Internet of Things.

In this book, we guide the reader through the numerous NFC applications that have evolved over the years or that are expected to come in the near future. We provide glimpses of a future where NFC technology has fully become part of our lives and serves as source of endless inspiration for anticipating the richness of the NFC world. This book also introduces a set of guidelines to design future NFC applications with a high user acceptance in consumer markets.

Technical information about all aspects of NFC technology is provided here. The areas covered range from basic concepts to research-grade material including future directions. This book captures the current state of NFC technology and serves as a source of comprehensive reference material on this subject. It has a total of 12 chapters authored by 50 experts from around the world. The targeted audience for the handbook includes professionals who are designers or planners for NFC systems, researchers (faculty members and graduate students), and those who would like to learn about this field. Although the book is not precisely a textbook, it can certainly be used as a textbook for graduate courses and research-oriented courses that deal with NFC. Any comments from readers will be highly appreciated.

Many people have contributed to this handbook in their unique ways. The first and foremost group that deserves immense gratitude is the group of highly talented and skilled researchers who contributed 12 chapters to this book. All of them have been extremely cooperative and professional. It has also been a pleasure to work with Rich O'Hanley and Stephanie Morkert of CRC Press, and we are extremely grateful for their support and professionalism. Our families have extended their unconditional love and strong support throughout this project, and they all deserve very special thanks.

Contributors

Manfred Aigner
Graz University of Technology
Styria, Austria

Elisabeth André
Augsburg University
Augsburg, Germany

B. Benyó
Sapienza University of Rome
(CATTID)
Budapest University of
Technology and Economics
Budapest, Hungary

Elisa Bertino
Qualcomm
San Diego, California

Abhilasha Bhargav-Spantzel
Intel
San Francisco, California

Thomas Bingel
Hochschule Darmstadt
University of Applied
Sciences
Darmstadt, Germany

Francisco Borrego-Jaraba
University of Cordoba
Cordoba, Spain

Pilar Castro Garrido
University of Cordoba
Cordoba, Spain

U. Biader Ceipidor
Sapienza University of Rome
(CATTID)
Rome, Italy
and
Budapest University of
Technology and Economics
Budapest, Hungary

Marta Cortés Orduña
University of Oulu
Oulu, Finland

Martin Feldhofer
Graz University of Technology
Styria, Austria

Miguel Ángel Gómez-Nieto
University of Cordoba
Cordoba, Spain

Eric Gressier-Soudan
Conservatoire National des Arts
   et Métiers
Paris, France

Rosa Iglesias
Ikerlan Technological Research
   Center
Mondragon, Spain

Felix Köbler
Technische Universität
   München
Munich, Germany

Philip Koene
Technische Universität
   München
Munich, Germany

Helmut Krcmar
Technische Universität
   München
Munich, Germany

Karin Leichtenstern
Augsburg University
Augsburg, Germany

Jan Marco Leimeister
Kassel University
Kassel, Germany

Irene Luque Ruiz
University of Cordoba
Cordoba, Spain

Michael Massoth
Hochschule Darmstadt
   University of Applied
   Sciences
Darmstadt, Germany

Guillermo Matas Miraz
University of Cordoba
Cordoba, Spain

C. M. Medaglia
Sapienza University of Rome
   (CATTID)
Rome, Italy
and
Budapest University of
   Technology and Economics
Budapest, Hungary

Philipp Menschner
Universität Kassel
Kassel, Germany

A. Moroni
Sapienza University of Rome
  (CATTID)
Rome, Italy
and
Budapest University of
  Technology and Economics
Budapest, Hungary

Romain Pellerin
Ubidreams
La Rochelle, France

Andreas Prinz
Universität Kassel
Kassel, Germany

Mikko Pyykkönen
University of Oulu
Oulu, Finland

Florian Resatsch
Servtag GmbH
Berlin, Germany

Jukka Riekki
University of Oulu
Oulu, Finland

Iván Sánchez Milara
University of Oulu
Oulu, Finland

Ning Shang
Qualcomm
San Diego, California

Michel Simatic
Institut Telecom
Telecom Sud Paris
Evry, France

Kevin Steuer Jr.
Qualcomm
San Diego, California

Juan Pedro Uribe
Ikerlan Technological Research
  Center
Mondragon, Spain

A. Vilmos
Sapienza University of Rome
  (CATTID)
Rome, Italy
and
Budapest University of
  Technology and Economics
Budapest, Hungary

1

SECURE MOBILE IDENTITY

Concepts and Protocols

ABHILASHA BHARGAV-SPANTZEL

Contents

Digital Identity Overview

The emerging information infrastructure connects remote parties worldwide through the use of large-scale networks, and through a diverse and complex set of software technologies. Activities in various domains, such as commerce, entertainment, scientific collaboration, healthcare, and so forth, are increasingly being carried out based on the use of remote resources and services. These resources and services are engaged at various levels within those domains. The interaction between different parties at remote locations may be (and sometimes should be) based on only little knowledge about each other.

To better support these activities and collaborations, information technology (IT) infrastructure and systems are needed that are more convenient to use. We expect, for example, that personal preferences and profiles of individuals will be readily available when shopping over the Internet or when running jobs on a computing grid, without requiring the individuals to repeatedly enter them. In such a scenario, digital identity management (IdM) technology is fundamental in customizing user experience, underpinning accountability in transactions, and complying with regulatory controls. For this technology to fully deploy its potential, it is crucial that strong *protection of digital identity* be achieved. IdM systems must assure that such information is not misused and individuals' privacy is guaranteed.

In this section, we describe the basic concepts related to digital identity followed by describing some key challenges in digital identity management and use. We also provide an overview of federated digital identity management systems.

Basic Concepts

Digital identity can be defined as the digital representation of the information known about a specific individual or organization. More specifically, our notion of digital identity refers to two different, not necessarily disjoint, concepts: nyms and partial identities. A *nym* gives an individual an identity under which to operate when interacting with other parties; an example of a nym is a login name or a pseudonym. Nyms can be strongly bound or linked to an individual, or be meaningful only in the context of a specific application domain. Weakly bound or unbound nyms are useful in contexts such as chat rooms and online games. *Partial identities* encompass a set of properties, such as name, birth date, credit card numbers, patient record number, which are referred to as *identity attributes* or *identifiers*, that are associated with individuals. We use an identity attribute as a synonym of an identifier. Each subset of identifiers represents the partial identity of the individual. Partial identities may or may not be bound to the human identity of one or more actual individuals.

It is important to note the issue of *identity ownership* as the identity attributes of individuals are stored and shared among various entities in IdM systems. By *owner of an identity attribute*, we mean the individual to whom this identity attribute is issued by a trusted authority or an individual who is authoritative with respect to the claiming of the identifier. In the former case, the trusted issuer of the identifier is also responsible for providing information about the *validity* of that identifier. The validity of an identifier encompasses several notions (some of which are derived from the field of data quality [24]). Examples of such notions are (1) correctness; the identifier is correct (possibly with respect to the real world); and (2) timeliness; that is, the identifier is up to date.

When talking about identifiers, it is also important to distinguish between weak and strong identifiers. A strong identifier uniquely identifies an individual in a population, whereas a weak identifier can be applied to many individuals in a population. Whether an identifier is strong or weak depends on the size of the population and the uniqueness of the identity attribute. The combination of multiple weak identifiers may lead to a unique identification [5,30]. Examples of strong identifiers are an individual's passport number or social security number (SSN). Weak identifiers are attributes such as citizenship, age, and gender. This distinction is significant because misuse of strong identifiers can have more serious consequences, such as identity theft, as compared to misuse of weak identifiers.

Our notion of identity verification deals with verifying that the identity attributes claimed by an individual are also owned by that individual. Identity verification is coupled with the concept of identity assurance. The notion of identity assurance deals with the confidence about the truth of the claims related to the identity of an individual. Successful identity verification with high assurance about an identifier claimed by an individual means that the identifier is considered valid and the verifier is confident that it is owned by that individual.

Strong and weak identity assurances exist regardless of the linkability of the identifier to the identity of the actual individual. Additionally, linkability among identifiers may exist with or without being bound (or linked) to the actual individual.

Example 1.1 Consider an individual whose real-world name is Bob Smith who has a digital pseudonym Homer07. In a digital interaction, when Homer07 claims to have SSN=123456789 and the verifier has strong assurance that the claim is correct (i.e., the SSN is valid and owned by the user Homer07) and linked to the real-world individual Bob Smith, then this corresponds to the case where one has strong identity assurance and strong linkability to the real-world individual.

Consider another scenario in which Homer07 claims to have citizenship=U.S.A. and the verifier does not know which real-world individual the claim belongs to, but at the same time, is confident that the claim is correct. Such a scenario corresponds to strong identity assurance and weak linkability. Notice that for a party to make a decision, such as in access control, linkability to a human identity of the actual individual is not always required.

In addition to the traditional identifiers, there also exist *biometric* identifiers that are increasingly included as an integral part of an individual's identity. Biometric verification occurs when an individual presents a biometric sample, and possibly some additional identifying data such as a password, which is then compared with the stored sample for that individual. Biometric verification provides some inherent advantages as compared to other nonbiometric identifiers because biometrics correspond to a direct evidence of the personal physical characteristics versus possession of secrets which can be potentially compromised. Moreover, most of the time biometric enrollment is executed in person and in controlled environments, making it reliable for subsequent use [23].

An interesting extension to the traditional identity attributes is the incorporation of the history of an individual's activities. The *transaction history-based identifiers* or *mobile identifiers* can be encoded as receipts from e-commerce or m-commerce transactions. This concept is especially relevant when considering mobile client systems such as cellular phones that are enabled with Near Field Communication (NFC)-type communication capabilities. This is elaborated further in the section titled "Mobile Identity Concepts."

Federated Identity Management Systems

The goal of identity management systems (IdMs) is to provide individuals with protected environments to share identities among organizations by managing individuals' identity attributes. Federations provide a controlled method by which federation members can provide more integrated and complete services to a qualified group of individuals. The members of a federation have trust relationships among themselves to share and use individuals' identity attributes.

Federations are usually composed of two main entities: IdPs that manage identities of individuals, and service providers (SPs) that offer services to registered individuals. In a typical federated IdM, the individual registers with his or her local IdP and is assigned a login name. Based on this information a registered individual can submit additional attributes and corresponding attribute release policies that are stored at the IdP. From then on, the IdP is contacted whenever the individual interacts with any SP in the federation and additional identity information is needed. The IdP is then in charge of sending the SP the submitted attributes of the individual in accordance with the attribute release policies. Note that there are several social, economic, and legal requirements to realize an IdM system. For example, the legal requirements would have to dictate how the contracts for transactions limited to the physical world get adopted when these transactions are performed electronically. Those nontechnical requirements are important to address when building an IdM system. In such *federated systems*, multiple IdPs are distributed and can store partial identity information of individuals, if required.

Mobile Identity Concepts

In this section, we present *history-based* or *mobile* identity attributes that are related to users' past transactions that can be used by users, together with other identity attributes, to perform identity verification and enabling SPs to make trust-based decisions concerning current transactions. One category of such systems is represented by *reputation systems* [18,14]. Several e-commerce SPs have built reputation systems so as to give a better idea of how trustworthy both the buyers and the sellers are. This is because the sellers are typically SPs but could also be users in a peer-to-peer (P2P) environment. Sellers benefit from the use of such systems because a good reputation score is likely to attract more customers. Similarly, buyers may qualify for better deals and services if they have a good reputation. However, most reputation systems have a major limitation in that the only information they maintain are scores, and they do not typically provide information about the actual transactions a seller or buyer has made. Therefore, it is important that trust be established also according to the transaction history–based attributes. Examples include the behavior of a given user with respect to compliance and usage while accessing cloud services such as storage or computational services or preferences related to what books the user bought, etc. Information about the history can be consulted to evaluate and manage the potential risks in a given transaction. Note that collection and management of this type of identity information needs to ensure that the *privacy* of the user is preserved throughout the identity management lifecycle.

Capturing and using transaction history for trust establishment entails addressing several challenges. First, there should be a privacy-preserving methodology to guarantee ownership of the history-based attributes on which the trust decisions are made. Moreover, in e-commerce applications, the transaction history of individuals includes their customer profile of transactions with several SPs and such transaction history needs to be accessed by various SPs, which may use heterogenous transaction history formats. In some existing real-world scenarios, the SPs store transaction history in such a way that makes it impossible for other SPs to use it. Therefore, the user cannot benefit from its past transactions. Additionally, there is a lack of *user control* on his or her transaction history. The transaction history is generally stored at the SP end, and the user may not be able to control who accesses this information. One solution is to introduce a third-party receipt management server. With such system, SPs can have access to the user's transaction history according to the users' permissions. The history-based attributes are encoded as receipts related to the past transactions. This model in turn can be used for m-commerce applications, where the importance of mobile identity and mobile devices can be well understood.

M-commerce applications allow users to conduct business and service transactions over portable wireless devices. One main challenge in an m-commerce transaction is providing users with the capability to prove their chosen past transactions to establish trust with the SPs. Such transaction history is used [31] to establish trust or provide history-based services. For example, users can prove their successful e-commerce transactions and qualify for rebates or discounts based on them. Because of the extensive commerce activities carried out by users both online and in person, it is in the interest of the users to be able to use their transaction history information in various types of m-commerce transactions.

Ideally, the transaction history of users should be easily accessible from any location, and encoded in a compact form that is acceptable from any service provider. An approach to addressing the requirements is to support electronic receipts encoding relevant information about past transactions, and use such receipts from mobile devices, such as cellular phones.

Supporting such type of receipts in mobile environments is very challenging. First, it is not trivial to ensure the security and privacy of the receipts. By using common technologies such as Bluetooth or RFIDs [3], it is possible to retrieve information from the phones without user consent, thus causing serious privacy concerns. A second issue is related to the storage and computational constraints of most cellular phones [3], which require ensuring the efficiency of the receipt protocols. Furthermore, in m-commerce applications it is desirable to integrate the transaction histories from online and in-person transactions, so as to maintain a comprehensive history. Some cellular phones can communicate with a store's information devices such as smart card reader, providing users alternative means for receiving and providing electronic receipts. Supporting these technologies in a secure fashion is nontrivial because of the threats emerging from the ubiquitous nature of mobile environments and the physical constraints of cellular phones.

To support mobile history-based transactions and cope with the aforementioned issues, users' identity attributes in an identity record (IdR) stored at the IdP can be leveraged. An NFC-enabled [3] cellular phone can be used to manage and use receipts for m-commerce transactions in a secure fashion. NFC is a standards-based, short-range (~15 cm) wireless connectivity technology that enables two-way interactions among electronic devices, allowing users to perform contactless transactions, access digital content, and connect electronic devices [3]. An NFC device embedded in the cellular phone is able to communicate not only with Internet via wireless connections but also with smart card readers. In addition, the cellular phone applications referred to as MIDlets can access the phone's tag for reading and writing data. Furthermore, such cellular phones can have extended memory that users can take advantage of for carrying the receipt information.

It would be very useful to establish and manage the history-based or mobile identity attributes using our mobile devices because of the nature of such attributes. Such devices can importantly be used to support the secure and privacy-preserving usage of this identity information. This could be used for both online as well as physical transactions. Moreover, cryptographic protocols, such as secret sharing and zero knowledge proof, can be used in a potentially vulnerable and constrained setting. Some of these considerations are now discussed.

Requirements

Online Transactions   There are several desired properties for an online transaction history management system.

Security Requirement. Security of the receipt protocols includes seven main properties:

1. **Correctness** means that if two honest parties successfully complete an e-commerce transaction, then the final receipt is constructed with the correct receipt attributes and is included in the database of users involved in the transaction.
2. **Integrity** refers to the tamperproofness of the constructed receipt. If any receipt attribute is modified, then it should be possible to detect the change.
3. **Single Submission** requires that the same receipt be not submitted more than once as two different receipts.
4. **Fairness** requires that the proof of delivery from the buyer and the proof of origin from the seller are available to the seller and buyer, respectively. Moreover, the protocol must be fail-safe, in that incomplete execution of the protocol must not result in a situation in which the proof of delivery is available to the seller but the proof of origin is not available to the buyer, or vice versa.
5. **Non-repudiation** for two-party protocols is twofold [28]: (a) non-repudiation of origin, that is, providing the buyer with irrefutable proof that the content received was the same as the one sent by the seller; (b) non-repudiation of delivery, that is, providing the seller with irrevocable proof that the content of the item or token received by the buyer was the same as the one sent by the seller.

6. **Stealing prevention** for receipts can be described using the following example: if a receipt R_{P_A} is issued to a user P_A, then P_B, who steals this receipt, should not be able to present R_{P_A} as its own receipt.

7. **Availability** of receipts is a critical requirement for mobile identity. If the transaction history is saved as cookies locally at the client machine, portability and hence the availability of such receipts, is hard to achieve. An infrastructure based on an identity management system should make sure that the receipt information is available to the online users.

Privacy Requirement. The privacy requirement for the receipt protocols consists of three main properties:

1. **User Consent** requires that the users be able to consent or agree to terms or conditions that may be associated with the disclosure and use of its receipt attributes. It is important that the user is given an opportunity to reject any disclosure of receipt information if required by the SP [21].

2. **Minimal Disclosure** requires that only the minimal piece of receipt information, as needed by the SP, is revealed.

3. **User Choice** allows the user to select parts of a receipt based on the information needed to carry on the current transactions.

Architectural Requirement. From the architectural perspective, there are three key considerations:

1. One desired property is that the system should be **easy to deploy** in current e-commerce systems with minimal extensions to the existing systems.

2. The **management** overhead imposed on individuals should be as low as possible so to assure **usability**.

3. Finally, the system should support **interoperability**, in that it should be possible to use the transaction history from one SP at another SP.

In-Person Transactions The use of receipts in off-line in-person transactions at physical SP locations using mobile phones also has some unique requirements. In the case of using history-based attributes in mobile devices, the user control and minimal disclosure properties are shown to be especially important [13] and should be supported. Moreover, the computational and resource constraints of such devices should also be considered to ensure *efficient* execution of the proposed protocols. In the following, we discuss relevant security and privacy requirements concerning the use of receipts on cellular phones in the context of in-person transactions.

Security of information stored in cellular phones is a critical factor for m-commerce transactions [13]. The following requirements concern the security of the receipt attributes and the receipt usage. We focus on protection requirements for the mobile device and do not consider other security requirements that may exist at the SP and registrar. First, the user should be able to prove ownership of the receipts recorded at the cellular phone. This requirement is related to non-repudiation, in that if ownership is proved, the user cannot deny having carried out the associated transaction. Second, the receipt attributes should correspond to the transaction for which the receipt was issued. Additionally, the integrity of the receipt attributes and other identity attributes stored at the receipt record should be assured. Finally, the secrets stored on the cellular phones should be trustworthy. Ensuring trustworthiness of secrets may require splitting them, or integrating independent parties or components on the cellular phones. It is important that the foregoing requirements be addressed at the same time to ensure flexibility and receipt portability.

To preserve *privacy*, it is important that users be able to maintain some level of control over their personal information and other identity attributes while conducting transactions with any SP. More specifically, we require that the retrieval of receipts should be driven by conditions based on the user preferences or the SP policies. In addition, the user should explicitly consent or be aware of the data being released from the cellular phone. Finally, the user should be able to disclose to the SP only the information strictly needed in order to satisfy the SP service policies, thus allowing minimal disclosure (Figure 1.1).

In the following we provide an intuitive example of possible m-commerce application of our protocols.

Example 1.2 Consider a user Alice who conducts an e-commerce transaction with SP e–Follets to buy a book for the price of $134.65 and then uploads the receipt corresponding to this transaction. As per the protocols provided,

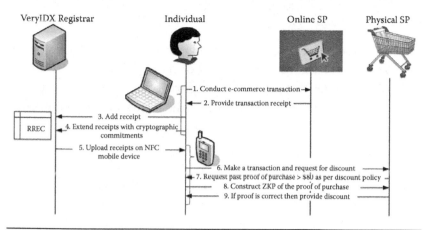

VeryIDX Registrar   Individual   Online SP   Physical SP

1. Conduct e-commerce transaction

2. Provide transaction receipt

3. Add receipt

RREC   4. Extend receipts with cryptographic commitments

5. Upload receipts on NFC mobile device

6. Make a transaction and request for discount

7. Request past proof of purchase > $80 as per discount policy

8. Construct ZKP of the proof of purchase

9. If proof is correct then provide discount

Figure 1.1   Example scenario of NFC cellular phone-based identity management.

Alice therefore uses her set of receipt attributes to create a cryptographic commitment on the price of the receipt (see [7]), which would be needed to create proof of ownership of the receipt and prove receipt attributes.

Alice uploads a subset R of the receipts at the registrar to her NFC cellular phone. Alice then decides to use her receipts at a physical SP shop Follets to qualify for a particular discount, which requires her to show that she has performed a (possibly electronic) transaction buying an item from the Follets or e–Follets for more than $80. The device retrieves the appropriate receipts based on the conditions specified by the SP. All the receipt information is not passed to the SP without the explicit consent of Alice. Moreover, to ensure minimal disclosure, Alice wishes to prove that the receipt from the e–Follets transaction is greater than $80 without showing the exact value in the clear. If the proof is correct, Alice qualifies for the discount offered by Follets.

Follets, on its turn, would like to make sure that the receipts are actually owned by the individual presenting the receipts. Using the cryptographic commitment stored in the cellular phone, which is signed by the registrar, Alice is able to prove to Follets that she owns the receipt that contains the receipt attributes being presented. The integrity of the receipt attributes involved can also be verified by Follets by checking the registrar's signature on the receipt attributes.

Sample Mobile Identity Use Protocols

In this section, we provide sample protocols in an identity management system that use the concepts of mobile identity presented earlier in this chapter and investigate the related security and privacy requirements. The focus is on the use of NFC-based devices that are used to perform m-commerce transactions.

Preliminary Notions

In this section, we provide background information. We begin with a brief overview of the key components of the cellular phone. We then illustrate the underlying cryptographic protocols adopted in the system, that is, the secret sharing protocols and the aggregate zero knowledge proof protocol.

NFC Cellular Phone Architecture   We employ the Nokia 6131 NFC cell phone (Ph^{NFC}) [3] for analyzing our portable receipts' protocols. We assume that the SPs have an NFC reader (denoted by NFC^{SP}_{reader}) that transmits and receives messages from the NFC cellular phone. The phone is integrated with an NFC device and thus contains both reader and writer to the embedded smart card and tags that directly communicate with the SP's reader. Ph^{NFC}'s components are shown in Figure 1.2.

As shown, the main software component for managing and using receipts is the MIDlet suite. The MIDlet suite consists of a Java Application Descriptor (JAD) and a MIDlet. A MIDlet (denoted by Ph^{mid}) is a Java program that runs on the Java Virtual Machine (JVM)-enabled mobile device, including a cellular phone. The JAD controls possible permissions the MIDlet gets. A Ph^{mid} is installed onto a phone and operates in a sandbox [29], so different MIDlets are isolated from each other, which increases security. The cellular phone consists of a secure

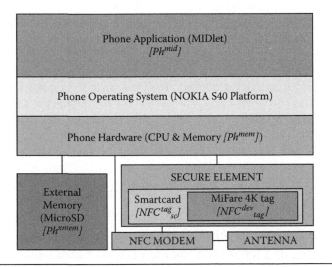

Figure 1.2   Sample NFC cellular phone components (based on Nokia Model).

element that can only be accessed by MIDlets signed by a trusted third party. The secure element consists of two main components, namely, the Mifare tag (NFC_{tag}^{dev}) and Smartcard (NFC_{sc}^{dev}). Since NFC_{tag}^{dev} and NFC_{sc}^{dev} are designed to contain sensitive information such as monetary information, only trusted MIDlet can access them.

Cryptographic Protocols

Shamir's Secret Sharing Scheme Secret sharing refers to methods for distributing a secret among a group of participants, each of whom is allocated a share of the secret. The secret sharing scheme by Shamir [25] splits a secret S into n partial secrets so that all n partial secrets are required to reconstruct S. Briefly, the scheme works as follows. First, $(n-1)$ random coefficients $\{a_1,\ldots,a_{k-1}\}$, with $a_0 = S$, are chosen. Then, a polynomial $f(x) = a_0 + a_1 x + a_2 x^2 + \cdots + a_{n-1} x^{n-1}$ is built. Based on $f(x)$, n shares are constructed. Each share is of the form $(i, f(i))$, where i denotes the input to the polynomial and $f(i)$ the output. Given n pairs, the coefficients of the polynomial can be evaluated using interpolation. The secret, that is, a_0, can thus be determined.

Aggregate Zero Knowledge Proof of Knowledge To be able to create proof on an identifier m, the user creates a Pedersen's commitment of the form $M = g_1^m h_1^r$, where r is a random value chosen by the user and g_1 and h_1 are public parameters of the registrar. This commitment is enrolled at the registrar. At enrollment, the registrar signs the commitment M to output $\sigma = M\chi$ as the signature, where χ is the secret key of the registrar.

The verification works as follows. Let $\sigma_1, \sigma_2, \ldots, \sigma_t$ be the signatures corresponding to the identifiers that need to be proved by the user to the SP. The user aggregates the signatures into $\sigma = \Pi_{i=1}^{t} \sigma_i$, where σ_i is the signature of committed value $M_i = g_1^{m_i} h_1^{r_i}$. It also computes $M = \prod_{i=1}^{t} M_i = g_1^{m_1 + \cdots + m_t} h_1^{r_1 + \cdots + r_t}$. The user sends σ, M, M_i, $1 \le i \le t$ to the verifier.

As a next step, the user and the verifier SP carry out the following ZKPK protocol*:

$$PK\{(\alpha, \beta): M = g_1^\alpha h_1^\beta, \alpha, \beta \in Z_q\}$$

---

* The convention is that Greek letters denote quantities the knowledge of which is being proved, whereas all the other parameters are sent to the verifier [9].

After the verifier SP accepts the zero knowledge proof of the commitments, it checks if the following verifications succeed:

$$M = \prod_{i=1}^{t} M_i \quad \text{and} \quad e(\sigma, g_2) = e(M, v)$$

where g_2 is a public parameter, v is the public key of the registrar, and e is a bilinear mapping. If the last step succeeds, then the verifier accepts the ZKPK of the signed commitments.

Protocols for Mobile Identity Management Using Cellular Phones

In this section, we present the approach to secure the secrets associated with the receipt attributes, using the Shamir secret sharing scheme and provide protocols to execute ZKPK proof on the cellular phone.

Sharing Secrets Used in Cellular Phone   Receipts are stored in the external memory Ph^{xmem} or other memory devices as given in Figure 1.2, depending on the user's choice and Ph^{NFC}'s storage availability. The protocols for receipt usage are independent of how these receipts are stored. The cryptographic secret and random value r used to create a Pedersen commitment are associated with a receipt attribute. Such a secret is used to create a ZKPK. We split this random value r into several different locations to increase security. Thus, only when all of the partial secrets can be merged can r be correctly reconstructed.

A key feature of using this approach is to allow the user to select different security levels by choosing different secret splitting levels. In the highest level of security, the secret required to do the identity verification transaction needs to be split in all four locations (see Figure 1.3).

When the ZKPK proofs corresponding to the receipt attributes need to be created, k secret shares are combined to create the original secret in the user's Ph^{NFC}. Shares are retrieved by the device according to their location. Specifically, the procedure to retrieve each secret share at the Ph^{NFC} is as follows:

1. s_1: Ph^{mem} is accessed directly by the secret combining function to retrieve s_1.
2. s_2: The user is required to input the secret PIN number p using the NFC cell phone's keypad. p', which is stored in Ph^{xmem}, is used to evaluate the second secret share $s_2 = p \oplus p'$.

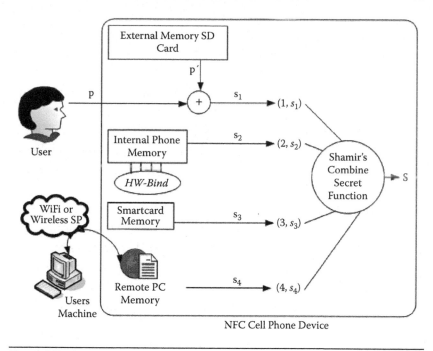

Figure 1.3 Secret splitting and combination.

3. s_3: The secret s_3, stored at the smart card, is read by the Ph^{NFC}.

4. s_4: The secret share s_4 is stored at the user's PC, which may not be present in the vicinity of the Ph^{NFC} when it is being used. As a first step, the cellular phone connects to a service that runs through the user's PC by remote procedure calls in order to retrieve the secret shares stored at the PC remotely, for example, the Windows NT Service. The activation requires the user to provide an authentication token. The authentication token may be automatically generated by the device if the user configures its device to not require explicit authentication token input when the service activation occurs. Once the token is successfully verified, the $NFC_{Internet}^{user}$ queries the PC "wallet" containing the secret share. After the secret share s_4 is retrieved, the $NFC_{Internet}^{user}$ deactivates the query service.

The secret shares are retrieved and combined by a new MIDlet Ph^{midc} that runs in a protected domain with restricted permissions. In this way, the secret reconstructed on the Ph^{midc} is isolated from access by other applications running on the Ph^{NFC} and can be used only as intended, without requiring that a copy be stored at the Ph^{NFC}'s memory.

Security, Privacy, and Usability Analysis Fundamental security and privacy properties such as minimal information disclosure and non-replay of proof of ownership are inherited by the employed ZKPK protocols. In the following, we instead focus on possible attacks on the MIDlets and on how our proposed protocols ensure user consent when releasing receipt attributes.

Software Security. The integrity of the MIDlet is ensured by using the trusted MIDlet suite for all the applications running our protocols. The trusted MIDlet suite is composed of signed code so as to ensure the integrity of the applications running on the Ph^{NFC}. An important aspect of MIDlet is that they run in a sandboxed environment [29], providing the necessary isolation of the memory usage between Ph^{mid} and Ph^{midc}.

One possible attack on the MIDlets can occur if an attacker intercepts the MIDlet application during proof creation, to either read the cryptographic secrets to compromise the confidentiality of such secrets, or to write to the MIDlet during proof creation, resulting in possibly incorrect proofs. To mitigate this attack, the information critical MIDlets (Ph^{midc}) should run in a restricted environment with no connectivity with external devices so that the attacker cannot use the excessive permissions or open ports, to access the Ph^{midc} to exploit any potential vulnerabilities. More specifically, one can consider two types of *trusted domains*. If the MIDlet needs to access cryptographic secrets such as Ph^{midc}, then it can be run in a restricted domain called the *critical protected domain*. This domain helps protect against the interception of possibly malicious programs to retrieve the secrets used in a given computation [3]. If other functionalities are needed, for example Ph^{mid} that needs connectivity with the NFC_{reader}^{SP}, then it runs in the *uncritical domain* that contains a set of permissions to access the Ph^{NFC}'s resources. The set of permissions is often called function group. An example of a function group assigned to Ph^{mid} is *Local Connectivity* that contains permissions related to connection via local ports, such as NFC, to handle the necessary communication with NFC_{reader}^{SP}.

A potential threat is of *malware attacks* that may try to retrieve the partial secrets from the cellular phone. In our context, splitting the secret into secure components that require signed code confers protection against malware attacks. For example, the secret share stored in the smartcard cannot be accessed by unsigned and potential malicious code.

Device Security. A cellular phone has a higher risk of *asset theft* as compared to ordinary desktop computers. The proposed use of secret sharing aims at protecting the secrets associated with sensitive identity information. However, this approach can be extended to any information that needs to be protected in a mobile device. The secret sharing mitigates the threat of asset theft in a threefold manner.

First, have a partial secret stored in a separate external memory that the user should keep detached from the cellular phone when the secret is not needed. If the partial secret from the external memory cannot be retrieved, then the original secret cannot be reconstructed. The second measure is to associate the partial secret s_2 with a user pin p (see Figure 1.3) that is not stored on the phone. The third measure taken is to require the online connectivity during the use of a split secret. When a cellular phone is stolen, the user can revoke the secrets stored in the device by calling the cellular phone service provider. At the time of verification during an m-commerce transaction, the verifier checks for such flag and can invalidate a transaction if the use of revoked secrets is identified.

Privacy. An attack on the user's privacy is an attempt by a malicious SP to access the receipt and other identity attributes from the user's mobile phone without the explicit consent of the individual. To prevent this threat, it is crucial to ensure user control [13]. An individual should provide explicit consent to every transaction or attribute receipt exchange. The user interface is critical in ensuring that the user is aware of which attributes are being revealed so that consent can be provided correctly.

Usability. The user interface is a very important aspect for ensuring the usability of identity management protocols using mobile devices. Also, portability makes it possible for users to have multiple devices (such as cellular phones and external storage devices) implementing the protocols, thus enabling user choice not only with respect to the attributes but also with respect to the device. Portability is achieved through adherence to standards and use of MIDlets for applications. MIDlets can be copied to other platforms and used to manage the receipt and other identity attributes collected at the RREC and the IdR.

The receipts can be stored and added depending on the memory capability of the Ph^{NFC} and the Ph^{xmem}. A Nokia 6131 NFC model can use up to 11 MB of Ph^{mem} and up to 2 GB of Ph^{xmem}. The unit size of a typical cryptographic receipt tuple containing one commitment is 4 KB. The Ph^{xmem} can hold up to 50,000 receipts for 2 GB memory, and the maximum number of receipts that can be stored into phone's internal memory is 2750, although this number will vary depending on the memory taken by other files in the phone, such as the multimedia files. Note that the individuals need to upload only a subset of receipts. The wallet contains the secrets needed for proof generation. With any modification of the secrets, the proof verification will fail, so integrity needs to be ensured. To ensure confidentiality of the secret residing in plain text in the Ph^{xmem}, the user can lock the memory card with a password. It is, however, not required that the password be reentered every time an access is made to the wallet file in the Ph^{xmem}.

Related Work

In this section, we discuss related work on the use of cellular phones for e- and m-commerce transactions involving identity attributes and other recent developments in mobile identity management initiatives.

With the advent of high-speed data networks and feature-rich mobile devices, the concept of *mobile wallet* [20,8] has gained importance. The initial efforts to combine digital cash and mobile telephones were under two distinct projects by CWI Amsterdam. One was on mobile device authentication, and the other on Chaum's online digital payment protocols [10]. In both cases, the idea was to connect to the bank and payee via the mobile networks, using the then newly introduced Global System for Mobile Communication (GSM) [22] mobile terminal as the payer's electronic wallet. Subsequently, as a part of the European CAFE e-commerce project [8], this idea was extended in a seminal work introducing the concept of wallets with observers [10] that enabled off-line digital cash and credentials to be used in commercial settings.

The project developed electronic wallet technology, where the transactions are performed via a short-range infrared link either directly with compliant cash registers and wallets held by other individuals or over the Internet to other SPs. Although the functionality of the CAFE wallet was never demonstrated in combination with cellular technology, the project was a significant step forward for mobile wallet technology. The observer is trusted by the credential issuer and protects the issuer's interests during off-line transactions. The observer restricts copying and uses the credentials on behalf of the issuer. An off-line transaction in this respect is a transaction where both the credential holder (payer) and the credential verifier (payee) do not connect to any auxiliary services. The purse is owned and trusted by the payer.

The wallets-with-observers approach was generalized in Reference 20, which also exploited the online mobility of the user's device, and the available wireless networks. They solved the multi-issuer problem in the original approach by having the mobile keep a single access credential corresponding to an entity called a localized credential keeper at the user's machine.

Another commercial example of a mobile wallet is the Valista mobile wallet [15], which supports functions such as secure payment transactions, personalization, and user identity verification. Such a system employs a provider-centric approach in which the wallets are hosted on a server (such as an IdP) and accessed from the user's mobile device. The wallets comply with the major security standards, such as Visa's Mobile 3-D Secure and MasterCard's Secure Payment Application (SPA). Other mobile identity management initiatives have gained importance with the rapid adoption of second-generation mobile telecommunication systems, leading to the growth of m-commerce [16,22]. Two specific factors critical in this domain are usability and trust. Several approaches to enhance usability of mobile devices have been proposed [12]. Trust on the device comprises several security and privacy properties such as confidentiality, integrity, user control, and minimal disclosure of the identity data stored on such devices. One approach to mobile IdM is based on GSM [22]. GSM-based IdM uses the GSM infrastructure and the Subscriber Identity Module (SIM) as the underlying platform. The use of GSM-based mobile IdM has several advantages, but the identity attributes managed are very limited and related to the SIM hardware or the GSM infrastructure. Identity attributes such as those involved in current IdM systems [2] are not supported. The sample approach presented earlier in this chapter could use the GSM infrastructure to provide history-based and other general identity attributes. There are also several privacy and trust issues using the proposed GSM model [22] that can be mitigated using our approach.

Summary

In this chapter, we have presented the concept of mobile identity attributes encoded as receipts related to online transaction histories of individuals and protocols to build and manage such attributes. It is possible that individuals' online activity be used to generate reliable identity information that can be managed and used as any other identity attributes to evaluate trust-relationship-based related properties such as reputation. We analyze how the receipts can be made portable, and used with mobile phone devices. In the mobile identity context, we stress the importance of having the protocols achieve the desired performance, usability, security, and privacy properties in the system. In essence, the concepts and models presented in this chapter provides a basis for a flexible and privacy-preserving methodology to use history-based attributes in an identity management framework.

There are specific assumptions to consider while employing the history-based protocols. The first is the participation of the SPs in issuing receipts to individuals as specified in the protocols. Second, at the time of verification, it is assumed that the SPs define policies based on such receipt attributes and attribute properties that can potentially be proven in zero knowledge.

The receipt protocols allow *user choice* when receipts are revealed to a given SP. Therefore, if a user does *not* provide a receipt, then it does not imply that the user did not execute a particular transaction. More specifically, the SPs do not gain knowledge of all potential transactions executed by the user, but instead those that the user chooses to reveal. Trust establishment based on the knowledge of all possible transactions of a given user will require additional mechanisms such as profiling.

An important aspect for successful deployment of protocols related to e-commerce is to analyze the constraints and requirements of the various e-commerce applications. For example, the secure electronic transaction (SET) [19] protocols that provided mechanisms to allow SPs to substitute a certificate for a user's credit card number failed to be implemented because of several practical considerations. A first consideration was with respect to the cost and complexity for SPs to support such protocols, especially given the presence of simpler alternatives such as SSL [27]. In addition, it was cumbersome to install client software and allow client-side certificate distribution. In our approach, we provide a flexible mechanism to allow various types of transactions depending on the capability and requirements of the system. Moreover, we show that there is minimal computational overhead and need for client software in our prototype implementations. However, to be practical, additional studies of human-computer interaction [6], market acceptance, and other business requirements are needed.

Bibliography

1. Federal Trade Commission fact sheet, Aberdeen group, identity theft: A $2 trillion criminal industry in 2005. http://www.ustreas.gov/offices/domestic-finance/financial-institution/c%ip/pdf/identity-theft-fact-sheet.pdf.
2. Liberty alliance project. http://www.projectliberty.org.
3. Near field communication forum. http://www.nfc-forum.org.
4. Consumer fraud and identity theft complaint data, January-December, 2005. http://www.consumer.gov/sentinel/pubs/Top10Fraud2005.pdf, 2006.
5. Ana I. Anton. Testimony before the House Committee on Ways and Means subcommittee on protecting the privacy of social security number from identity theft. *US Public Policy Committee of the Association for Computing Machinery*, 1:1–19, June 2007.
6. Abhilasha Bhargav-Spantzel, Anna C. Squicciarini, Matthew Young, and Elisa Bertino. Privacy requirements in identity management solutions. In *Twelfth International Conference on Human-Computer Interaction*, Lisbon, Portugal, 2007.
7. Abhilasha Bhargav-Spantzel, Jungha Woo, and Elisa Bertino. Receipt management transaction history based trust establishment. In *DIM '07: Proceedings of the 2007 Workshop on Digital Identity Management*, Fairfax, Virginia, 2007.
8. Jean-Paul Boly, Antoon Bosselaers, Ronald Cramer, Rolf Michelsen, Stig Fr. Mjolsnes, Frank Muller, Torben P. Pedersen, Birgit Pfitzmann, Peter de Rooij, Berry Schoenmakers, Matthias Schunter, Luc Vallee, and Michael Waidner. The ESPRIT project CAFE - high security digital payment systems. In *ESORICS*, pages 217–230, 1994.

9. Jan Camenisch and Markus Stadler. Efficient group signature schemes for large groups. In *Advances in Cryptology—CRYPTO '97*, pages 410–424, 1997.

10. David Chaum. Security without identification: transaction systems to make big brother obsolete. *Communications of the ACM*, 28(10):1030–1044, 1985.

11. Rachna Dhamija and Doug Tygar, J. The battle against phishing: Dynamic security skins. In *SOUPS '05: Proceedings of the 2005 Symposium on Usable Privacy and Security*, pages 77–88, New York, 2005. ACM Press.

12. Alan Dix, Tom Rodden, Nigel Davies, Jonathan Trevor, Adrian Friday, and Kevin Palfreyman. Exploiting space and location as a design framework for interactive mobile systems. *ACM Transactions on Computer Human Interaction*, 7(3):285–321, 2000.

13. Sastry Duri, Marco Gruteser, Xuan Liu, Paul Moskowitz, Ronald Perez, Moninder Singh, and Jung-Mu Tang. Framework for security and privacy in automotive telematics. In *WMC '02: Proceedings of the 2nd International Workshop on Mobile Commerce*, pages 25–32, New York, 2002. ACM Press.

14. Minaxi Gupta, Paul Judge, and Mostafa Ammar. A reputation system for peer-to-peer networks. In *NOSSDAV '03: Proceedings of the 13th International Workshop on Network and Operating Systems Support for Digital Audio and Video*, pages 144–152, New York, 2003. ACM Press.

15. Denis Hennessy. The value of the mobile wallet. White Paper, December 2003. www.valista.com.

16. Uwe Jendricke, Michael Kreutzer, and Alf Zugenmaier. Mobile identity management. Technical Report 178, Institut für Informatik, Universität Freiburg, October 2002.

17. Otto Kolsi and Teemupekka Virtanen. Midp 2.0 security enhancements. In *HICSS '04: Proceedings of the Proceedings of the 37th Annual Hawaii International Conference on System Sciences (HICSS'04)—Track 9*, page 90287.3, Washington, DC, 2004. IEEE Computer Society.

18. Kwei-Jay Lin, Haiyin Lu, Tao Yu, and Chia en Tai. A reputation and trust management broker framework for web applications. In *EEE '05: Proceedings of the 2005 IEEE International Conference on e-Technology, e-Commerce and e-Service*, pages 262–269, Washington, DC, 2005. IEEE Computer Society.

19. Larry Loeb. *Secure Electronic Transactions: Introduction and Technical Reference*. Artech House, Norwood, MA, 1998.

20. Stig F. Mjolsnes and Chunming Rong. Localized credentials for server assisted mobile wallet. *ICCNMC'01: International Conference on Computer Networks and Mobile Computing*, 00:203, 2001.

21. Andrew S. Patrick and Steve Kenny. From privacy legislation to interface design: Implementing information privacy in human-computer interfaces. In *Proceedings of Privacy Enhancing Technologies Workshop (PET2003)*, *LNCS 2760*, 2003.

22. Kai Rannenberg. Identity management in mobile cellular networks and related applications. Information Security Technical Report 9, Johann Wolfgang Goethe University Frankfurt, January 2004.
23. SC 37 Secretariat. Text of FCD 19795-2, biometric performance testing and reporting—part 2: Testing methodologies for technology and scenario evaluation. In *ISO/IEC JTC 1/SC 37*. ANSI, Geneva, Switzerland, 2006.
24. Monica Scannapieco, Paolo Missier, and Carlo Batini. Data quality at a glance. *Datenbank-Spektrum*, 14:6–14, 2005.
25. Adi Shamir. How to share a secret. *Communications of the ACM*, 22(11):612–613, 1979.
26. Mercan Topkara, Ashish Kamra, Mikhail J. Atallah, and Cristina Nita-Rotaru. ViWiD: Visible watermarking based defense against phishing. In *International Workshop of Digital Watermarking (IWDW)*, volume 3710, pages 470–483, 2005.
27. David Wagner and Bruce Schneier. Analysis of the ssl 3.0 protocol. In *WOEC'96: Proceedings of the Second USENIX Workshop on Electronic Commerce*, pages 4–4, Berkeley, CA, 1996. USENIX Association.
28. Guilin Wang. Generic non-repudiation protocols supporting transparent off-line TTP. *Journal of Computer Security*, 14(5):441–467, 2006.
29. Alexander Wolfe. Toolkit: Java is Jumpin'. *Queue*, 1(10):16–19, 2004.
30. David Woodruff and Jessica Staddon. Private inference control. In *CCS'04: Proceedings of the 11th ACM Conference on Computer and Communications Security*, pages 188–197, New York, 2004. ACM Press.
31. Giorgos Zacharia, Alexandros Moukas, and Pattie Maes. Collaborative reputation mechanisms in electronic marketplaces. In *HICSS*, Maui, Hawaii, 1999.

2

SECURITY IN NFC

MARTIN FELDHOFER AND MANFRED AIGNER

Contents

Introduction to Contactless NFC and RFID Systems

NFC is a combination of contactless communication technologies. On the one hand, NFC enables direct interaction between mobile devices such as phones or cameras, and on the other hand it allows access to RFID-tagged items and contactless smart cards. Interacting with passive devices, the NFC handheld acts as a contactless reader. It provides the electromagnetic field that powers the tag or smart card and triggers the communication. We expect that in the future many items will contain such tags. Communication with our mobile phone with those tags or cards will happen on a regular basis. Depending on the application, protection of the communication with those tags is necessary to prevent, for example, end-user privacy violation or tag cloning attacks. While the contactless communication standard for interaction with smart cards (ISO-14443) provides a security layer, the current communication standard for low-cost RFID tags (ISO-15693) is still unprotected. NFC allows communication of both kinds of protocol types. This chapter discusses protection of the communication between NFC devices and low-cost RFID tags. Current sales figures for this kind of tag lead us to the assumption that in the near future objects with such tags will be common.

In general, RFID is a means of automatically capturing data from objects. The technology is relatively old, with its first uses in military applications in the 1940s. A massive growth in the industry started in the late 1990s with the possibility of fabricating low-cost RFID tags in high volume. NFC allows access to those tags not only from dedicated reader devices, but from devices we carry with us every day (e.g., a mobile phone or camera). It will represent, therefore, the major interface between our daily life and tagged objects.

The basic principle of RFID technology is that a device called a reader or interrogator, in our case a mobile phone with an NFC interface, provides an RF field. It transmits data to the tag, which is also known as a transponder, by modulating the field. The tag is an integrated circuit that is attached to an antenna. It receives requests and transmits data to the reader. Unfortunately, the working principle is only elementary at first glance. When considering the vast number of functional parameters and different configuration possibilities, the technology becomes very complex. However, this chapter provides an overview of RFID technology in general but focuses on the topics that are relevant for the implementation of passive RFID systems in combination with NFC technology.

The following introduction is presented from the perspective of a tag manufacturer. In a passive RFID system, the reader provides the power for operation of tags via an electromagnetic field. The bidirectional communication allows identification and sending of data by applying different interaction techniques depending on the used frequency range.

The definitive reference is the *RFID Handbook* [9] from Finkenzeller, which deals with a multitude of topics such as the physical basics, the device construction, and applications of RFID systems. Even German literature is available. An interesting book is from Fleisch and Mattern [10], which is a collection of papers concerning RFID applications. Another good book is from Kern [20], who gives a nice introduction to the technological aspects.

In the following text, the description of the working principles gives an overview of what we mean by passive NFC/RFID tags. The system parameters such as operating frequency, modulation technique, and power supply will be examined, and functional requirements such as bulk reading and reading range will be elaborated on. A section on applications will showcase the numerous applications of RFID technology in practice. Because regulation and standardization are important to ensure interoperability, an entire section is dedicated to these issues. In our coverage, we will focus on HF technology, since this is the frequency range compatible with NFC. Nevertheless, where characteristics are similar, we will also refer to UHF RFID tags.

Operating Principle of Contactless Systems

The working principles of RFID systems mainly depend on the used frequency range. Nevertheless, the common fundamentals of all passive RFID systems should be explained by the HF system shown in Figure 2.1. NFC operates in the HF range (13.56 MHz). The RFID reader (or a

of sight between the reader device and the tag. Many processes can be automated, and it is possible to write data to the tag and modify them. These properties help prevent goods getting lost during transportation and shelves in stores becoming empty. An RFID-enabled logistic chain can help prevent expiry of perishable goods and allows detection of counterfeiting of products.

The applications that evolved during the last few years are much more interesting as a simple barcode replacement. The following applications provide an overview of where processes have been optimized using RFID technology. In these applications, we do not distinguish between different types of RFID tags (labels or cards) and the employed frequency range (13.56 MHz in the case of NFC).

Supply-Chain Management Supply-chain management (SCM) is the process of organizing the movement and storage of materials and goods. It is one of the main applications of RFID technology. The biggest advantage of an RFID-based supply chain is the high visibility of goods, which allows efficient controlling possibilities in real time. Nowadays, most RFID systems use tags at the pallet level due to the high costs of the tags. In the near future, item-level tagging will probably become more popular. The establishment of an open supply chain from the production of a product to the end of its lifetime improves many business processes. During production, the availability of the required materials is improved. Transportation can be optimized, and stock keeping can be done more efficiently. Even the customers can benefit from the attached RFID tag on a product if warranty issues can be handled by using RFID technology. NFC-equipped mobile phones play an important role in this respect, because they enable customers to interact with the tags. The majority of SCM applications use UHF technology, but some market segments prefer an HF tag because they are less susceptible to interference with metal and liquids.

Inventory Management Inventory management can be a subdivision of supply-chain management, but for many applications it is a discipline in its own right. Many library systems (e.g., at TU Graz) use RFID tags in their books. This allows finding books in the shelves much faster and eases the borrowing process. Returning the books also becomes very simple. The inventory management of pharmaceuticals makes such products more safe. The expiration date can be checked automatically, some protection against counterfeiting can be achieved, and the tags can be used for disposal of unused medicaments. NFC allows integration of the end consumer and pushes the border of applications beyond the companies' facilities.

Tracking and Tracing   Tracking and tracing is an issue that can be optimized by the use of RFID systems. Many automatic toll collection systems use either active or passive RFID tags for tracking purposes. RFID tags for tracking of patients in hospitals improves documentation and avoids mistakes of wrong medication and therapies. Tracing of documents is applied in governmental institutions with increased transparency and efficiency. At sport events, where thousands of people have to be measured quickly, RFID tags are used to register the start and the end time of individuals.

Logistics   In logistic applications, RFID systems are used to automate processes that are typically very time intensive for humans. The transport of luggage at airports is an example of RFID technology highly increasing the efficiency. Other logistic applications are the postal service, waste disposal, and the transport of military goods.

Access Control   A very popular application of RFID systems is access control. Instead of cheap RFID tags, mostly cryptography-enhanced RFID cards (smart cards) are used. Many public transport systems and companies all over the world use such cards or NFC-enabled devices for access control. Hands-free ticketing systems are deployed in skiing areas and at public events. Nearly every modern car has a remote keyless entry system and an immobilizer system based on RFID technology.

E-Passport   The introduction of electronic passports should help prevent forgery of passports and should help to increase throughput at checkpoints. Most e-passports have a contactless smart card that operates at 13.56 MHz according to the ISO/IEC 14443 RFID standard [14]. The passport provides authentication of the owner by a digital signature calculated over the personal data stored on the chip. In order to prevent the possibility of tracking people, an access control mechanism is implemented that only provides access to the data if the reader is authenticated by the passport.

Anti-Theft Systems   Anti-theft systems based on RFID technology use many different types of tags. The prevention of shoplifting is mostly done by electronic article surveillance (EAS) systems, which only indicate whether there is a tag in the range of the reader or not. In comparison, car immobilizers often use contactless smart card technology, while in libraries rewritable RFID tags are utilized that indicate if a book has been legally borrowed or if it has been stolen.

Proof of Origin   Proof of origin is not only important for electronic passports. Many products such as car parts, tires, and other branded goods are targets for counterfeiting. A possible solution is to attach an RFID tag to these products. Instead of a simple identification using the UID of an RFID tag, cryptographic authentication is suggested as proof of origin. Throughout this work, this is one of the main target applications for which we implement cryptography-enhanced tags.

Implementation of Passively Powered RFID Tags

We presented a general overview of the functionality of RFID systems in the previous section; we now turn to the implementation of passively powered RFID tags.

The architecture of an HF tag is similar to that of a contactless smart card or memory card. The main difference is the power consumption. While contactless cards have a typical power consumption in the milliwatts range, passive tags must not dissipate more than a few microwatts. This results in a much larger operating range of up to 1.5 m compared to a few centimeters for contactless cards, when operated with a long-range reader. Another distinctive feature is the costs of the devices. Contactless memory cards with a price around €1 are typically used in applications such as access control where the devices are frequently used. In contrast to such memory cards, RFID tags are much cheaper. According to one renowned company, the medium-term goal is to produce five-cent tags. This would allow using RFID tags in many more applications, even at the item level. Such low costs can only be achieved when the ICs are very tiny, the tag production process is highly optimized, and billions of tags are produced per year.

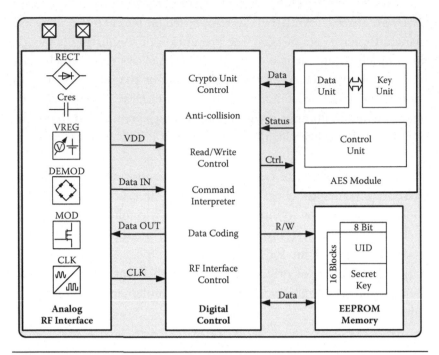

Figure 2.2   Architecture of a passive HF RFID tag.

Architecture of Passive NFC Tags   The construction of a passive RFID tag is in general very simple. It mainly consists of an IC that is attached to an antenna. The antenna is used to draw energy for operation of the IC out of the RF field of a reader. Furthermore, the antenna is necessary to allow bidirectional communication between reader and tag. The main challenge is the cost-efficient production of tags that satisfy the rigorous constraints on power consumption and chip area. The architecture of a passive HF RFID tag is depicted in Figure 2.2. The IC is attached to a coil antenna that is placed on a label. The main components of the IC are the analog front-end, the digital control unit, the nonvolatile memory, and optionally modules for additional functionality such as cryptography or sensors. The analog front-end provides the interface between the two antenna pins and the digital control unit. The first task is to provide a stable power supply for the whole IC. Depending on the frequency range, different rectifier circuits (e.g., bridge rectifiers, charge-pump rectifiers) are used for this purpose. A capacitor of a few 100 pF is used to smooth the resulting voltage. Furthermore, it stores the energy that can be used when the field is not available due to a modulation by the reader or when computational intensive operations have to be performed. Depending on the current distance from the reader, the induced voltage highly differs. Hence, a voltage regulator, which provides a constant supply voltage for the digital part of the circuit, is necessary. In order to indicate to the digital controller that the supply voltage is stable (e.g., during power-up when the tag enters the field), a power-on-reset signal has to be provided. The second task of the analog front-end is to establish a bidirectional communication link between reader and tag. Demodulation of mostly ASK-modulated signals of the reader works similarly for the HF and the UHF frequency range. The rectified signal is filtered, and after an envelope detector a comparator is used to digitize the signal. Tag-to-reader communication works differently for HF and UHF. In HF tags, load modulation is used where a resistor parallel to the antenna is switched on and off according to the sent data. For UHF tags, the mechanism is called *backscattering*. Here, a capacitor or a resistor are switched by a transistor. The power reflected by the tag is varied, and hence data can be transmitted. The third task of the analog front-end is to generate a clock signal for the digital circuit.

The digital control unit is the central part of the tag and determines its functionality. The incoming digital data stream from the demodulator is time continuous. It has to be sampled with an appropriate clock frequency, and data have to be decoded. When the data frames are available, the command interpreter carries out the appropriate operation. This can be either to read or write to the memory, perform the anticollision sequence, or execute any other command. The example in Figure 2.2 shows that the digital controller could also steer a cryptographic module in the RFID tag. The controller reads and writes data to the module and starts cryptographic operations. Finally, the digital controller is responsible for sending responses to the reader. The transmitted data have to be coded, which results in a digital output stream that is used as input for the modulator circuit. The nonvolatile memory has to store data that have to be retained when the tag is not supplied by a reader. Such data are, for example, the unique identifier, data about the tagged product, and the secret key of the cryptographic operation. It is also possible to lock diverse memory regions against overwriting. This protects, for instance, the unique identifier of an RFID tag from getting manipulated. Typically, the nonvolatile memory is implemented as an EEPROM with sizes from several bytes up to a few kilobytes. Today, cryptographic modules are only included in contactless smart cards but not in passive RFID tags. We assume that in the near future also, low-cost RFID tags with cryptographic features will appear on the market.

Manufacturing Process of Passive NFC Tags An RFID tag usually consists of an integrated circuit that has been attached to a flexible substrate. On this substrate an antenna of a specific form has been printed. The resulting inlay is then assembled between a printed label and its adhesive backing film, yielding a passive RFID tag. During the manifold test procedures, the tag is programmed with its unique identifier. Finally, it is attached to a pallet, case, or item for being used in an RFID application. A short overview of the production process of a complete RFID tag will now be described. The first step after designing the integrated circuit is the production in silicon. Semiconductor chips are manufactured on so-called wafers, which are thin plates of pure silicon having typically a diameter of 15 cm to 30 cm. Many dies, which have sizes between 0.25 mm^2 and 1 mm^2, are fabricated on a single wafer. For example, approximately 8000 dies for ISO/IEC 15693 RFID tags are available on a 15 cm wafer using a relatively old CMOS process technology. A first test is executed directly on the wafer because the earlier an error is detected, the lower are the costs. For the wafer test, many dies are connected in parallel by using needle probes. In addition to a functional test, the ICs are preconfigured by writing the unique identifier to the memory. This memory region is then set to "read only" to prevent changing the UID afterward. The wafer is subsequently sawed along the boundaries of the dies, and the chips that have been marked as defective during the wafer test are rejected. The naked dies are packaged to protect the IC against environmental influences and mechanical stress. Often, the assembled chips are put on an intermediate carrier, the so-called strap. A strap consists of two metal plates that are connected to the antenna pins of the IC for later attachment to an antenna. This is a standardized placement methodology where on a reel, which is about 16 cm broad, many tags can be placed. The final test by the IC manufacturer is done on large testing machines on a reel-to-reel basis. With such equipment, testing up to 60K devices per hour is possible. The design of the antenna has a great influence on the performance of the tag. The typical shape of HF tags are coil antennas. All types of antennas typically use one of the following manufacturing processes. They are either printed using conductive ink, etched on a metal foil, stamped, or vapor-deposited. Commonly used materials are either copper, aluminum, or silver. Assembly is the last production step of an RFID tag. A standard RFID tag uses an adhesive carrier on which the die (or the strap on which the IC has been placed for easier handling) is connected to the antenna. A plastic or paper label is put on top to protect the components of the tag. The final test of the assembled RFID tag is to check the functionality, the conformance, the performance, and the interoperability. The functional test simply decides whether the tag works or not. Conformance or compliance testing checks if the device conforms to a specific standard. The conformance test methods itself are also standardized (e.g., ISO/IEC 18047-3). The tests defined in ISO/IEC 18046 allow the performance verification of tags under certain boundary conditions of the measurement, such as orientation and range. Interoperability tests check the possibility of operating with different reader devices. "Golden" readers are used for different applications. As a final step, before the end users attach the tags to their goods, custom data are written to the tag. The presented manufacturing process allows separating the costs of an RFID tag into three equally large parts. One third corresponds to the costs for IC manufacturing. The antenna production also contributes one third to the overall tag costs. Finally, the assembly process is also responsible for one third of the expenses. The consequence of this cost estimation for designing cryptographic hardware in the course of this work is that additional chip area should be kept small.

Introduction to Security for Passive NFC Tags

In the early years of passive RFID technology, security was not considered an important topic. Passive RFID tags were considered as replacements for barcodes for applications in the supply chain. While security measures for smart card technology were transferred to contactless smart cards, the primary goal for design of RFID tags was cost saving, and therefore minimal functionality was included in the tags themselves. The advantages over barcodes, such as improved reading distance, bulk reading, and elimination of the necessity for direct line of sight, were addressed. The resulting security concerns (tag cloning, eavesdropping, etc.) were not taken into consideration. This chapter outlines the development of the security topic in the context of RFID applications. We focus on passive RFID technology for longer reading distance without discussing the development of contactless smart card technology. When operating with NFC devices, basically the same security requirements arise. The fact that reading range is restricted to centimeters when operated with an NFC device does not mean an eavesdropping attacker cannot intercept from a greater distance. Attackers may take tags home and read them with any other reader. We do therefore think that the security requirements from pure RFID applications and those that use NFC devices as readers are identical.

The Privacy Discussion

Soon after the presentation of first ideas for applications of RFID technology, an intense public discussion about the technology's privacy implications emerged. This public discussion was mainly driven by privacy activists and consumer advocates and was based on many factoids.

The RFID-tag-producing industry ignored the discussion for a rather long time. They did not react to the statements but claimed that RFID technology does not have any privacy implications, as long as RFID tags do not hold personal data. This statement is true only for a very specific selection of applications. In the closed-loop supply-chain scenario, this argument is true as long as it is guaranteed that the tag is removed or killed when the tagged item is handed over to the end consumer. Over the years, the discussion became active, sometimes with strange arguments such as RFID chips are "the mark of the beast" [19] from Biblical prophesies or wrong statements such as RFID tags can be read from satellites.

In fact, the ongoing discussions led to a situation that turned out to be critical for development of the technology. One famous incident was the announcement of a huge RFID project by Benetton. Philips Semiconductors published that Benetton was about to order a very large number of tags to introduce item-level tagging for optimization of their garment supply chain. In fact, the project was in a very early stage and far from introduction into the productive system, but vague ideas for use after the point of sale were already discussed (e.g., the RFID-enabled washing machine).

No item-level tagging was considered in the first stage of the project; pallet-level tagging was planned as a start. In the planned scenario, no privacy problems would appear for the end customers because the purchased goods would not carry RFID tags. Only after the first stage of the project was considered successful would item-level tagging be considered later on. Directly after publication of the story by Philips, Benetton was flooded with a very high number of complaints and requests for statements, but they were not prepared for such a situation at that moment.

No privacy impact evaluation or anything similar was done at that stage of the project. A group called CASPIAN (Consumers Against Supermarket Privacy Invasion and Numbering) organized a boycott of Benetton products and got a good publicity in the press. Due to the high pressure, Benetton had to withdraw the project, and obviously also the order for RFID tags. So far, the project has not been implemented, or at least no public information about the results and reasons to stop the project are available. It is obvious that Benetton had to stop the project and kept away from RFID technology for the following years as a result of the story.

In 2006, the European Commission started discussions to solve the privacy issue in RFID technology. A series of public RFID consultations were organized together with a platform for public discussion.

In May 2009 the "Commission Recommendation on the implementation of privacy and data protection principles in applications supported by radio-frequency identification" was published. This recommendation provides a basis for the RFID industry to develop and implement applications and gives end consumers the confidence that their privacy will not be violated by the technology or the applications. According to these regulations, future RFID applications that involve processing of personal data require a Privacy Impact Assessment (PIA) before roll-out.

Requirements for this PIA are currently defined by all stakeholders in close cooperation with the European Commission. The RFID industry has also reacted to the ongoing discussion. NXP Semiconductors (successor of Philips Semiconductors) has already announced their new generation of RFID tags with built-in support for consumer privacy protection. The privacy topic and discussion is now addressed, and instead of claiming that the technology does not have a privacy impact, a new generation of tag products has come up that supports consumer privacy.

Added Value Due to Security Services

Often, security requirements for RFID systems are treated as "nonfunctional requirements." This approach does not consider benefits to the system's functionality triggered by embedded security features.

We suggest that security functionality be treated as a service that enables new application areas for a system or that raises the value of the system due to its protection. This approach facilitates the justification of the additional costs due to the protection. When the argument demonstrates benefits, the associated costs can be compared with the assumed rise in income.

We claim that a service-oriented approach is necessary for proper development of secure RFID technology. So far, in many discussions on secure RFID, the drawbacks of security (additional costs, reduced throughput, more complicated interaction) were highlighted. We suggest focusing on the benefits a protected system can provide in future discussions.

In the Internet, the introduction of SSL, or the secure version of http—shttp—enabled a multitude of new applications that were not feasible before (online banking, eGovernment, eCommerce, . . .). This can happen as soon as RFID systems can provide a proper level of security.

In the following, we provide a nonexhaustive description of services for RFID systems that can be achieved when tags with cryptographic capabilities are used. Some of the described services can be established by other means, for example, by data mining techniques applied to the data stored on servers of the EPC network, but we think that introduction of cryptographic functionality in tags is the more efficient choice.

Security Service: Proof of Origin or Anticloning   Tags with encryption functionality and a stored secret key can implement a challenge-response authentication protocol. To authenticate a tag, a reader queries the tag to receive its unique identifier (UID). Then the reader sends a random number as challenge to the tag, which encrypts the challenge under its secret key. The encryption result is sent back to the reader. Using the claimed UID, the reader can either retrieve a secret key from a database to perform the same encryption or send the UID and the challenge to an authentication server, which stores the key for the transmitted tag UID. The server calculates the expected tag response and sends it to the reader. When the expected challenge is computed, the reader verifies the authenticity of the tag by comparing the expected tag response with the one received from the tag.

When such a tag is attached to a product, this mechanism can be used as protection against product cloning. Companies can personalize the tags in their products with secret keys. An authentication server with public reading access can provide the necessary authentication data to customers or to any other party who intends to check the origin. Using an NFC mobile phone, an end user can check the authenticity of goods on the basis of their embedded RFID tags.

The tags can still be used for traditional RFID applications, for example, for supply-chain management or RFID-assisted inventory, without using the authentication feature.

Security Service: Data Integrity Protection for Data from Tags In current RFID systems, the transponders do not hold a lot of data, but their main functionality is to transmit their UID. Future RFID systems will include tags with increased memory; therefore, readers will store data on tags, or tags will generate data by themselves (e.g., tags with a temperature or light sensor). NFC readers may read the data from the tags.

When such tags are read and the data is relevant input for an application, it is important to check the integrity of the data coming from the tag. An illustrative example is a temperature sensor tag used to record the permanent temperature characteristic of an object in a cold chain. If somebody could change the temperature logging data coming from the tag, the logging would be useless. A sleazy operator of a cold chain could easily alter the logging data using his mobile phone with NFC interface to hide discontinuities under his responsibility. For food products, the goods will go bad earlier. For pharmaceutical or chemical products, such an interruption can cause severe damage with relatively high compensation claims.

In the case of unprotected data on tags, an attacker can harm the system by simply changing the logging data of correctly reposited goods, which would force a cold chain operator to dispose of nonexpired goods due to the forged logging data.

Whenever tags can get in the hands of nontrusted parties, protection against illicit changes of the data carried by the tags is necessary. In future applications, where tags are read by nontrusted readers and the acquired data is sent via the Internet to a back-end service, it is necessary to protect the data from changes by any node between the tag and the point where the data from the tag is processed.

Security Service: Access Control for the Tag's Memory or Commands   When data is written to tags, or the configuration of tags can be changed by a reader outside a trusted environment (e.g., execution of the kill command), it is necessary to provide protection from unwanted access.

For current tags, the write commands or the execution of the kill command is protected by passwords (access password or kill password in Gen2 tags). This is only a weak protection due to the rather short length of the password (32 bits) and the possibility of eavesdropping and reusing the password.

With the kill command, reuse is not a critical issue (after killing the tag, they do not operate anymore) if every tag is configured with a different kill password. This leads to complex password management.

Especially when write access to tags should be performed outside protected areas, another form of protection is necessary. A tag can request proper authentication of the reader before it grants access to critical memory areas or commands. The reader authentication can be performed on the basis of cryptographic protocols so that an eavesdropping attacker does not gain any information about the stored secret key. Reuse of the intercepted communication is then not enough to gain access to the tag's memory. If only authenticated and trusted readers get permission to write data to the tag or to execute critical commands, a broad band of attacks is repelled.

In future open and distributed RFID systems, it will be additionally necessary to hand over control of the tag's content. When a tagged object changes its owner, it is also necessary to perform a so-called transfer of ownership of the access rights to the tag's memory. After the transfer, only the new owner can alter the tag's data and configuration. This can be performed by exchange of the cryptographic keys used in the reader authentication process. Such transfer-of-ownership protocols will require interaction of the tag, the old, and the new owner.

Security Service: Protected or Encrypted Transactions between Reader and Tag   To repel eavesdropping attacks on the wireless channel between tag and reader, the exchanged data can be encrypted. A prerequisite for useful encryption is authentication, because the party who encrypts the data needs to decide which key should be used. Even if a session key for the encryption is generated by so-called "key agreement schemes," both parties need to authenticate to avoid man-in-the-middle attacks. Although NFC is designed for short-range access, it is still necessary to protect the communication between tags and NFC device. Attackers may use equipment that is more sensitive than the reception circuitry of the NFC device; therefore, the eavesdropping distance is much higher than the operational range.

A correctly established encrypted channel between authenticated parties (tags and readers) also prevents illicit tracking and tracing of RFID tags. No information about the content is revealed to an eavesdropper, but the exchanged data seems like a stream of random data.

Privacy protection for the owner of the tagged object is not the only motivation for encryption of RFID transactions. The data transferred between tags and readers during an inventory process can be an interesting target for industrial espionage. Such attacks can also be avoided by storing encrypted data on the tags, which is then sent during a transaction. This has the advantage that the tag itself does not require encryption functionality but simply stores and sends encrypted data. In such cases, the data delivered by the tag is static; this means an eavesdropper may identify tags by messages that were eavesdropped on previously. This can be a problem for the final application, when tracking or tracing of tags should be impossible. Encryption functionality on the tag itself can avoid such problems. Often, NFC is discussed for payment application. Even if the transferred amount of such payments is small for a single transaction, an automated attack after successful eavesdropping of transaction secrets may cause considerable damage.

Security Flaws in Existing Products

Already today, RFID tags are used in critical security applications. In accordance with our suggestion, most of these applications rely on proprietary cryptographic solutions. To achieve the power requirements for tags, dedicated algorithms that promised to result in a low-power design were developed. In many approaches, the used key length does not fulfill established cryptographic standards, and therefore brute force attacks become feasible.

In the following paragraphs, we provide information about recently published attacks on RFID applications with secure tags. We provide information about the attacks and try to pinpoint the weakness of the system that made the attacks feasible.

Texas Instruments: Digital Signal Transponder (DST)   The DST can implement a cryptographic challenge-response protocol to authenticate tags to a reader. The proprietary custom block cipher DST-40 is used as the underlying cryptographic primitive. Those tags were designed for use in an electronic car immobilizer system, with protected passive tags embedded in the car key.

At the time when the tags were designed, it was assumed that a 40-bit key provides sufficient protection against car theft. Designers of the system claim that at the time when the tags were developed, it was technically not possible to implement protection measures with longer bit lengths with the available silicon technology. In a later application, the RFID tags from the car keys were used to authenticate clients at payment terminals of gas stations (Speedpass by ExxonMobil).

In Reference 1, a group of students and researchers from Johns Hopkins University present a hack that exploits serious vulnerabilities of the tag. They used information from a presentation held by one of the developers of DST-40 [17] to reverse-engineer the proprietary and confidential encryption primitive. Knowing the algorithm, they implemented an FPGA-based key-search engine that is able to search the secret key for a given challenge-response pair within an hour. Once they know the secret key, they can copy it to a tag-emulation device to authenticate illicitly to the electronic car immobilizer system or to a payment terminal. While it is possible to check for duplicate tags in the online system of the payment application and to defeat this attack at this level, the electronic protection of the car key is completely broken by the attack. The researchers produced and published videos in which they showed that they could make a purchase on an electronic terminal and start a car using their emulation device. The necessary budget for the devices used in the attack was below US $10,000.

This incident was the first published attack on a protected RFID application. It received worldwide press coverage. The researchers demonstrated the weaknesses of an established system with a large number of terminals and tags in the field. With the introduction of electronic car immobilizers with passive RFID tags, the number of stolen cars has significantly decreased. Therefore, one can claim that the application is still useful because the effort needed to break it—together with the problem of circumventing traditional protection measures against car theft—is still too high for car thieves. Nevertheless, the decision to use the same protection in payment systems was a very critical one, and was taken probably years after the development of the tags themselves.

This successful attack shows that application of proprietary custom algorithms for protection is very critical. As soon as details about the implementation become public, attacks become probable. For future tag products, we must not scale the security level of the tags to a specific application, since it might be the case that the same tags are used in a different application with very different security requirements, as happened in this case. It was already clear at the time when the DST was developed that payment systems require a key length that is longer than 40 bits.

NXP: The MIFARE™ Incident   MIFARE describes a series of contactless smart card products from NXP.[1] Although MIFARE products employ contactless smart card technology, we decided to include the incident in this section, especially because the low-cost branch of MIFARE cards, which are intended for one-way tickets, have characteristics similar (minimal chip area) to RFID tags. Different versions of MIFARE with different security levels are available; the one we refer to here in this section is MIFARE Classic, which uses the proprietary CRYPTO-1 algorithm as security primitive. MIFARE was also licensed to other chip producers. Currently, MIFARE is the market leader and the de facto standard for contactless ticketing system. It was also discussed as protection for RFID applications.

The functionality of CRYPTO-1 was reverse-engineered with home equipment by graduate students. During the 24th Chaos Communication Congress in December 2007, they presented their results. Later on, they published their analysis results as a research paper [24]. During their work, not only was the functionality of the proprietary and undisclosed algorithm revealed, but also other grave weaknesses of the design were detected. Soon after disclosure of CRYPTO-1, an extremely efficient attack that reveals the secret key within some minutes followed.[22]

At the same time, a second academic team was investigating the security of MIFARE cards. Soon after the first presentation of successful attacks on MIFARE, researchers from Radbound University published a series of papers [11,2], where successful attacks on MIFARE cards and their applications were presented. After these publications, it was clear that the protection of MIFARE Classic cards was completely broken.

It is beyond doubt that MIFARE products are a commercial success. When the successful hacks were published, newer versions with stronger, and above all, standardized cryptographic features were already available as products. Nevertheless, MIFARE Classic cards are still used in many applications. The incidents show again the risk of using proprietary cryptographic primitives. As soon as the secret algorithm was disclosed, successful attacks became feasible.

Keeloq^{TM}: A Successful Attack Using Side-Channel Analysis Keeloq is available as product for active (battery-powered) and passive transponder devices. The main application area is battery-powered (active) Keeloq transponders for remote keyless entry systems. Due to the lightweight implementation and the similarity of keyless entry applications and possible applications of secure passive RFID tags, recent attacks on Keeloq are discussed here. In early 2008, a cryptanalytic attack on the Keeloq encryption algorithm was presented [12]. This attack requires exhaustive computation but is technically feasible. The impact on practical implementations was considered minor due to the computational effort necessary for an attack. Shortly after publication of the cryptanalytic attack, researchers from the University of

Bochum analyzed several "high secure" keyless entry systems with Keeloq protection by application of Side-Channel Analysis (SCA) methods. This class of attacks uses the physical characteristics of the encryption device, for example, the power consumption of the device during operation, to reveal the secret key. After analysis of several products, they completely broke the system [4]. Now they are able to reveal the secret key of remote devices and—even more critical—the manufacturer key, which allows generation of valid key values for cloned remote devices. In a follow-up paper, they explain how to reveal the secret key of a remote device after eavesdropping only two ciphertexts from the device [18]. Additionally, this attack allows prevention of access for legitimate devices, while allowing the illegal attacker to still enter. These attacks pose a serious threat for all installed systems protected by Keeloq. This incident is very similar to the previously described attacks on RFID systems. Compared to the value protected by the devices, it is possible to break them with rather less effort. It is interesting that this attack relies on an attacking method that was not known when the system was developed. The first public paper of SCA was published in 1996 [21], at a time when Keeloq was already available as a product. It is important to consider this point for future developments. The lifetime of tags in the field is rather long, compared to software products. Again, a proprietary algorithm was used that turned out to be vulnerable after profound investigation by independent researchers.

Future Attacks The academic community's interest in breaking RFID systems has just started. We expect more successful attacks on established systems in the future. Since successful attacks guarantee very high press coverage, they can boost academic careers. At least such attacks raise the international visibility of the concerned researchers significantly.

There is no reason to assume that other protected RFID tag solutions provide better protection. The design of currently available tags dates back to a time when the necessary protection level for the applications was underestimated and when silicon technology for tags did not allow better protection. The threat of implementation attacks such as SCA is an especially serious one for current secure RFID products.

Some people see this hacking activity of academic researchers as destructive, because it damages the business case of companies. These companies have invested money to bring their product to the market; a published attack can reduce sales and cause consequent financial damage. Nevertheless, we have to consider that such systems are designed to protect other values, and clients pay for proper protection. In case the protection is not as good as claimed, due to advances in technology or due to flaws in protection, this should be publicly known.

One can assume that criminal organizations also analyze protection measures, and possibly have more financial support than academic researchers. In case criminal parties find a security hole, they would not publish their findings but would try to profit from the flaw for as long as nobody else is aware of it. Such an incident would damage the customer who runs the inadequately protected application. The resulting legal case poses therefore high risk also for the producer of the protected system. The loss can then be higher than after publication of a security hole. Considering this, it is easier to see the benefit and importance of public analysis of protection measures.

Security Extensions for Existing RFID Protocols

In order to extend the functionality of today's RFID systems, it has to be ensured that security enhancements can be integrated into existing RFID protocols. In RFID systems, the limited computing power and the low-power constraints of the tags require special considerations concerning the used protocols. In addition to the available bandwidth for data transmission, attention should be paid to compatibility with existing standards such as ISO/IEC 18000-3 [15], ISO/IEC 18092 [13], or UHF EPC Gen2 [5].

In the following text, we will show a security-layer extension for ISO/IEC 18000-3, which is an enhancement of joint work in [7,8]. Although the ISO/IEC 18000-3 standard defines an HF system using a carrier frequency of 13.56 MHz, the concepts can be applied for many RFID standards in different frequency ranges (e.g., EPC Gen2 UHF). The related RFID standard is commonly used for item management, whereas it defines the operating conditions under which RFID tags are operated.

The proposed security layer is based on symmetric challenge-response authentication protocols as shown in the previous section. The Advanced Encryption Standard (AES) [22] has been selected as the cryptographic primitive. The reason for this is that AES is suitable for implementation in passive RFID tags. Nearly all presented security protocols need some kind of random numbers. These nonces (number used once) are not necessarily true random numbers. It is important that the numbers are not predictable and that they are only used once. The implementation in an RFID tag could be a pseudo-random number generator (PRNG), or often even a counter is sufficient.

Security Layer for ISO/IEC 18000-3

A typical RFID standard describes different sets of commands. In addition to the mandatory commands, which all tags must support, custom commands can be specified. The mechanism for the anticollision sequence has to be implemented by every tag. The mandatory inventory command is used for this purpose. The anticollision algorithm retrieves the UIDs of all tags in the reader's environment and works in the following way. The reader sends an initial inventory command. All tags in the environment respond with their UID. If only one tag answers the request, the UID can be retrieved by the reader, and all subsequent commands can be addressed solely to that tag using the UID. If two or more tags answer, a collision occurs. This can be detected by the reader. The reader then uses a modified inventory request where it adds a part of the tag's UID to the request. Only tags that match with part of the UID are allowed to answer. Once the UID of one tag is identified, the reader sends an addressed stay-quiet command to the tag, which does not attend to the anticollision sequence any more. This method is used until there are no more collisions, and all tags within the environment are identified. Adding an authentication command to the ISO/IEC 18000-3 standard works by the definition of a custom command. The challenge-response protocol ideally fits the overall request-response protocol of RFID systems. When authenticating a tag, the reader sends a challenge that is addressed for a single tag by its UID, which has been retrieved previously. The tag answers according to the presented authentication protocol with the

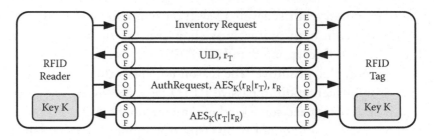

Figure 2.3   Mutual challenge-response authentication protocol based on AES.

encryption result. Sometimes it is useful to separate the authentication command into two distinct commands. First, a challenge is sent to the tag that sends no immediate reply. Then a second request asks for the encrypted result of the tag (see the interleaved protocol in the next subsection). Figure 2.3 depicts the mutual challenge-response authentication protocol that extends the ISO/IEC 18000-3 standard. First, the reader sends an inventory command within the command delimiters start-of-frame (SOF) and end-of-frame (EOF).

The tag answers with its UID and a random challenge r_T. Provided that no collision occurs, the reader sends an authentication request to the tag. By sending $AES_K(r_R \mid r_T)$ and a challenge r_R, the reader authenticates to the tag who can verify this encryption result. The tag provides authentication by encrypting the challenges in reverse order, $AES_K(r_T \mid r_R)$, and sending it back to the reader who can verify this response. Note that the common secret key K has to be stored in the reader (back-end database) and written to the tag during the personalization phase. A detailed analysis of various protocols to achieve unilateral authentication of tag and reader, mutual authentication with or without anonymity, etc., can be found in Reference 3. Furthermore, this publication provides information about the required times and the number of tags that can be identified per second using the presented challenge-response authentication protocols that are based on AES.

Interleaved Authentication Protocol

Integrating the proposed symmetric challenge-response authentication protocol into the ISO/IEC 18000-3 standard requires some additional considerations. Due to the fierce low-power constraints, the internal clock frequency of the RFID tag must be reduced from

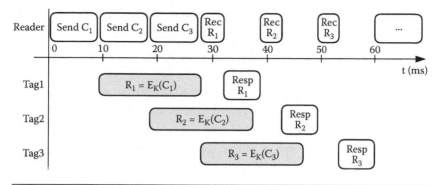

Figure 2.4 Interleaved authentication protocol for securing RFID tags.

the carrier frequency of 13.56 MHz to around 106 kHz. The under-lying standard demands that a response of the tag must follow 320 μs after a request from the reader. Otherwise, the tag has to stay quiet. The resulting number of clock cycles is 32. This is not suffi-cient for encrypting a challenge using a standardized algorithm such as the AES under the given low-power requirements. The solution to this problem is to modify the authentication protocol as shown in Figure 2.4. The challenges and the responses are interleaved with each other.

Normally, there are a lot of RFID tags to be authenticated in the environment of a reader. After retrieving all UIDs of the tags using the inventory request and the anticollision sequence, the reader sends a challenge C_1 to *Tag1*. This tag immediately starts the encryption of the challenge without sending any response. In the meantime, the reader sends further challenges to the tags *Tag2* and *Tag3*. These tags also start encrypting their challenges after reception. After finish-ing the encryption of $E_K(C_1)$, *Tag1* waits for the request to send the encrypted value R_1 back to the reader. When the reader has sent, for example, three challenges, it sends a request for receiving the response from *Tag1*. The received value R_1 is verified by encrypting the chal-lenge C_1 and comparing the result with the received value. The two other unanswered challenges are received using the same method. Then the reader starts from the beginning authenticating all other tags in the environment. This protocol has been evaluated using high-level models of the RFID communication channel and is a proof of concept for a security extension in an existing RFID protocol. The interleav-ing method has the advantage that the time for the tag's encryption is not determined by the maximum time to answer. In the discussed example, where three tags have been interleaved, the available time is 18 ms. This means that an encryption can take more than 1900 clock cycles at a clock frequency of 106 kHz. It is possible that up to 50 tags can be authenticated per second. If another algorithm (e.g., public-key encryption) would require more time for calculation, it is possible to enhance the interleaving to many more tags. The total time for authenticating all tags is solely determined by the amount of data that has to be transmitted. However, it is important to consider that in the worst case (only one tag is in the field), the time for authentication increases according to the time for the cryptographic primitive. An intelligent RFID reader can decide depending on the number of tags and the required time for calculation of the algorithm how many tags are requested in an interleaved manner to optimize the throughput.

Example of Secure Supply Chain Using RFID Technology

For a proof of origin of goods, authentication is essential. The following example shows how symmetric authentication of RFID tags can be used for an open supply-chain application. The proposed solution is taken from Reference 7. The application in Figure 2.5 shows a typical supply-chain scenario as it could be used in the pharmaceutical industry where e-pedigrees are required. HF tags are discussed in this marked segment as best-fitting technology; such tags are compatible with NFC.

The involved entities are the manufacturer, the freight forwarder, the drugstore, and the consumer. The mechanism of distributing the goods is simplified because in a real-world application there would be some intermediate trader between the manufacturer and the

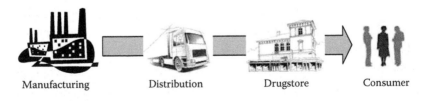

Manufacturing   Distribution   Drugstore   Consumer

Figure 2.5 Overview of supply-chain application.

drugstore. Nevertheless, the presented concepts can easily be adapted for further entities.

The first task of the manufacturer in an RFID-based supply chain is to stick the RFID tags on the products to be identified. Additionally, the personalization of the tag has to be done. This consists of writing the UID and the secret key to the tag's memory. The backend database has to store this information for later reference. It is also possible that the manufacturer operates some kind of web services to make general information and security information of the medicaments publicly available. During distribution of goods, the inventory functionality of the tags is used by the forwarding agents. This helps make the shipment more efficient and less error-prone. In addition, the customs declaration can be automated, and the authenticity of the drugs can easily be verified. In the drugstore, the RFID functionality can be used to improve stockkeeping. This includes inventory management, efficient ordering to prevent overstock, recalling defective products, and proof of authenticity to detect counterfeit products. The consumer should also be able to check the proof of origin and get additional information about the present batch of drugs. NFC-equipped mobile phones would allow the interaction of the consumer.

The verification procedures in the authentication process can be executed by various entities. The executing logistics companies are not necessarily trustworthy. Therefore, they are not in possession of the secret key. Verification of authenticity of a tag does not need the secret key. Valid challenge-response pairs are sufficient to prove identity. These pairs can easily be provided by the manufacturer, for example, as a web service. The applied mechanism depends on whether the verifying party has online access or whether the verification with the phone in the shop occurs off-line.

Off-Line Verification During transportation and distribution of goods, the verification entity might not have permanent online access to the authenticated server (e.g., during transport by sea). Therefore, a certain number of challenge-response pairs can be stored on the verification device in advance (in our case, a PDA with an RFID reader). This is the only online transaction that is necessary during the preparation phase. Figure 2.6 shows this first protocol phase, where

Figure 2.6   Collection of challenge-response pairs in advance.

the PDA transmits a request including information about the tags to authenticate to the authenticated server.

The server responds with pairs of challenges and the corresponding responses. As one product family might use more than one authentication key, more than one response is sent for a challenge in such cases. Every challenge-response pair is indexed with the corresponding UID of a certain product.

The verification phase in the field works as shown in Figure 2.7. After performing the anticollision sequence to retrieve the tag's UID, the reader chooses one of the stored challenge-response pairs and sends a challenge within an authentication request to the tag.

The response of the tag can be verified by the reader by comparing the stored result with the response. When the number of stored challenge-response pairs is too small, the problem of replay attacks may arise. Therefore, it is important either to use every challenge only once or to send different authentication requests to the tag to check whether the tag performs the encryption or if it only stores challenge-response pairs that have been recorded previously.

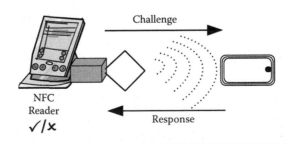

Figure 2.7   Off-line verification of tag authenticity.

Summary and Conclusion

The importance of security in RFID and NFC systems has been motivated thoroughly. The following main threats to RFID systems have been worked out. Unprotected tags are susceptible to counterfeiting because the tag's unique identifier can be cloned easily. The UID is transmitted unprotected over the air interface during the tag identification procedure. Additionally, the transmission of confidential data between the NFC device and the tag is also affected by eavesdropping. Another issue is tracking and tracing of people who wear or carry tagged products. Unprotected access to the tag's memory allows unwanted modifications of data stored on the tag.

Published attacks against RFID products, which are still used for contactless payment and for access control, showed that proprietary security algorithms are not suitable for protection of such systems. The susceptible communication link and the different computing capabilities of reader and tags make the implementation of security solutions for RFID systems difficult. The cost-intensive implementation of crypto primitives for passive RFID tags is especially challenging. Nevertheless, to achieve the security assurances confidentiality, authentication, and non-repudiation, cryptographic primitives with a high security level are necessary.

When NFC devices communicate with RFID tags, strong cryptographic algorithms and protocols that have been standardized are required for many possible applications. The advantages of standardized solutions are that the appropriate key size and the used primitives have been scrutinized by many crypto experts.

Recent research results have shown that implementation of standardized cryptographic functionality is technically feasible on low-cost tags. Cryptographic functionality of tags needs to be carefully considered during protocol design to avoid security holes or unnecessary overhead of secure operations.

Integrating security enhancements into existing RFID protocols is important to achieve compatibility of diverse systems. A security layer for interaction with NFC phones is necessary and can be established when tags feature cryptographic functionality. A security-enhanced supply-chain application has shown how counterfeiting can be prevented by authentication of RFID tags.

A comprehensive survey on recent technical research on the problems of privacy and security in RFID systems has been presented by Juels [16]. Further interesting security aspects of RFID systems have been presented by the National Institute of Standards and Technology (NIST) in Reference 23. Similar to the ARTICLE 29 Data Protection Working Party of the European Commission [6], they propose to use state-of-the-art security measures to protect RFID systems.

Endnotes

1 http://www.nxp.com/
2 http://www.cs.virginia.edu/kn5f/Mifare.Cryptanalysis.htm

References

1. Steve Bono, Matthew Green, Adam Stubblefield, Ari Juels, Avi Rubin, and Michael Szydlo. Security analysis of a cryptographically-enabled RFID device. In *USENIX Security Symposium, Baltimore, Maryland, USA, July–August, 2005, Proceedings*, pages 1–16. USENIX, 2005.
2. G. de Koning Gans, J.-H. Hoepman, and F.D. Garcia. A practical attack on the MIFARE Classic. In *8th Smart Card Research and Advanced Application Workshop (CARDIS 2008)*, volume 5189 of *Lecture Notes in Computer Science*, pages 267–282. Springer, 2008.
3. Sandra Dominikus, Elisabeth Oswald, and Martin Feldhofer. Practical Security for RFID: Strong Authentication Protocols. In Patrick Horster, editor, *Proceedings of D.A.C.H. Mobility 2006, October 17–18, 2006, Graz, Austria*, pages 187–200. Syssec, 2006.
4. Thomas Eisenbarth, Timo Kasper, Amir Moradi, Christof Paar, Mahmoud Salma-sizadeh, and Mohammad T. Manzuri Shalmani. On the power of power analysis in the real world: A complete break of the keeloq code hopping scheme. In *CRYPTO*, pages 203–220, 2008.
5. EPCglobal. EPC Radio-Frequency Identity Protocols Class-1 Generation-2 UHF RFID Protocol for Communications at 860 MHz—960 MHz Version 1.0.9, January 2005. Available online at http://www.epcglobalinc.org/.
6. European Commission—ARTICLE 29 Data Protection Working Party. Working document on data protection issues related to RFID technology, January 2005. Available online at http://www.europa.eu.int/comm/privacy.
7. Martin Feldhofer, Manfred Aigner, and Sandra Dominikus. An Application of RFID Tags using Secure Symmetric Authentication. In Panagiotis Georgiadis, Stefanos Gritzalis, and Giannis F. Marias, editors, *1st International Workshop on Security, Privacy and Trust in Pervasive and Ubiquitous Computing—SecPerU 2005, Santorini Island, Greece, July 14, 2005, Proceedings*, pages 43–49. Diavlos Publications, July 2005. In conjunction with the IEEE ICPS'05.

8. Martin Feldhofer, Sandra Dominikus, and Johannes Wolkerstorfer. Strong Authentication for RFID Systems using the AES Algorithm. In Marc Joye and Jean-Jacques Quisquater, editors, *Cryptographic Hardware and Embedded Systems—CHES 2004, 6th International Workshop, Cambridge, MA, USA, August 11–13, 2004, Proceedings*, volume 3156 of *Lecture Notes in Computer Science*, pages 357–370. Springer, August 2004.

9. Klaus Finkenzeller. *RFID-Handbook*. 2nd edition, Carl Hanser Verlag, Munich, April 2003.

10. Elgar Fleisch and Friedemann Mattern. *Das Internet der Dinge: Ubiquitous Computing und RFID in der Praxis:Visionen, Technologien, Anwendungen, Handlungsanleitungen.* Springer-Verlag, New York, 2005. ISBN 3540240039.

11. Flavio D. Garcia, Gerhard de Koning Gans, Ruben Muijrers, Peter van Rossum, Roel Verdult, Ronny Wichers Schreur, and Bart Jacobs. Dismantling MIFARE classic. In S. Jajodia and J. Lopez, editors, *13th European Symposium on Research in Computer Security (ESORICS 2008)*, volume 5283 of *Lecture Notes in Computer Science*, pages 97–114. Springer, Heidelberg, 2008.

12. Sebastiaan Indesteege, Nathan Keller, Orr Dunkelman, Eli Biham, and Bart Preneel. A practical attack on keeloq. In Nigel P. Smart, editor, *EUROCRYPT*, volume 4965 of *Lecture Notes in Computer Science*, pages 1–18. Springer, Heidelberg 2008.

13. International Organisation for Standardization (ISO). ISO/IEC 18092: Information technology—Telecommunications and information exchange between systems—Near Field Communication—Interface and Protocol, April 2004.

14. International Organization for Standardization (ISO). ISO/IEC 14443: Identification cards—Contactless integrated circuit(s) cards—Proximity cards, 2000.

15. International Organization for Standardization (ISO). ISO/IEC 18000-3: Information technology AIDC techniques—RFID for item management—Part 3: Parameters for air interface communications at 13.56 MHz, March 2004.

16. Ari Juels. RFID security and privacy: A research survey. *IEEE Journal on Selected Areas in Communications*, 24(2):381–394, February 2006.

17. Ulrich Kaiser. Universal immobilizer crypto engine. Guest presentation during Fourth Conference on the Advanced Encryption Standard (AES), 2004. The presentation slides were available via the conference web; they were removed after publication of the attack.

18. Markus Kasper, Timo Kasper, Amir Moradi, and Christof Paar. Breaking keeloq in a flash: On extracting keys at lightning speed. In Bart Preneel, editor, *Progress in Cryptology—AFRICACRYPT 2009, Second International Conference on Cryptology in Africa, Gammarth, Tunisia, June 21–25, 2009. Proceedings*, volume 5580 of *Lecture Notes in Computer Science*, pages 403–420. Springer, Heidelberg, 2009.

19. Liz McIntyre and Katherine Albrecht. *The Spychips Threat: Why Christians Should Resist RFID and Electronic Surveillance*. Nelson Current, Nashville, Tennessee, 2006.

20. Christian Kern. *Anwendungen von RFID-Systemen*. Springer-Verlag, Berlin, 2007.
21. Paul C. Kocher. Timing attacks on implementations of Diffie-Hellman, RSA, DSS, and other systems. In Neal Koblitz, editor, *Advances in Cryptology—CRYPTO '96, 16th Annual International Cryptology Conference, Santa Barbara, California, August 18–22, 1996, Proceedings*, number 1109 in *Lecture Notes in Computer Science*, pages 104–113. Springer, Heidelberg, 1996.
22. National Institute of Standards and Technology (NIST). FIPS-197: Advanced encryption standard, November 2001. Available online at http://www.itl.nist.gov/fipspubs/.
23. National Institute of Standards and Technology (NIST). Guidelines for securing radio frequency identification (RFID) systems, April 2007. Available online at http://csrc.nist.gov/publications/nistpubs/800-98/SP800-98_RFID-2007.pdf.
24. Karsten Nohl, David Evans, Starbug Starbug, and Henryk Plötz. Reverse-engineering a cryptographic RFID tag. In *SS'08: Proceedings of the 17th Conference on Security Symposium*, pages 185–193, Berkeley, CA, 2008. USENIX Association.

3

NFC Applications with an All-in-One Device

ROSA IGLESIAS AND JUAN PEDRO URIBE

Contents

NFC is a short-range wireless communication technology that has emerged from the convergence of contactless identification technologies such as RFID and networking technologies such as Bluetooth and Wi-Fi. NFC does not try to replace these technologies, but to coexist with them, complementing, enhancing, or bringing new experiences to users.

In this chapter, we guide the reader through the numerous NFC applications that have evolved over the years or that are expected to come in the near future. First, a brief summary of the main strong points of NFC over other wireless technologies is provided. After that, we look at NFC-enabled mobile phones as the goose that lays golden eggs. Finally, before we begin to explore the applications, this chapter includes a basic explanation of the three modes of operation of NFC technology, illustrating the operational basis of NFC applications.

We hope this chapter provides glimpses of a future in which NFC technology has fully become a part of our life. Also, we hope it serves as a source of endless inspiration for anticipating the richness of the NFC world.

Advantages of NFC

Initially, the physical and behavioral aspects of NFC technology are described, and afterward, the companies and institutions that have contributed to the growth of this technology.

NFC: Physical and Behavioral Aspects

The advantages of NFC compared to other technologies are attracting significant interest for developing new applications. For example, NFC operates only when two devices are in close proximity (generally, around 4–5 cm), unlike RFID technology, which can read RFID tags over a longer distance. As a result, since its range is quite short and consequently intercepting signals is difficult, NFC-based transmissions are inherently secure. The NFC operating range is up to 20 cm, whereas Bluetooth's range varies from around 1 m to 100 m, and Wi-Fi can reach up to 300 m in an open environment.

Furthermore, unlike Bluetooth and Wi-Fi, the connection between two NFC devices is established at once, without involving either user intervention or device pairing in the initial discovery phase or other configuration settings. Apart from its zero configuration, NFC consumes significantly less power than these technologies, and it works even when one of the devices is not powered. This can be achieved thanks to the use of inductive coupling, which is typical of RFID. Thus, when an NFC-enabled mobile phone is turned off, NFC is still working in a passive way. An NFC device is considered active if is provided with a power supply and can transfer energy and data from one device to the other.

Compared to RFID, which is its closest brother, NFC can function as either RFID cards or readers. What is more, since it is compliant with ISO/IEC 14443, NFC can communicate and exchange data with existing RFID equipment such as the most widely extended NXP's Mifare (Mifare 2010) and Sony's FeliCa (Felica 2010). Similarly to RFID, NFC tags can be very small (0.5 mm^2) and thus can be integrated in any device, have memory capacity (from kilobytes to over 1 Mb), and low price.

From the early days of its existence, all these features have contributed to forecasts of promising NFC applications, yet apart from the aforementioned advantage, two aesthetic advantages that can dramatically contribute to the growth of NFC are still missing. So far, we have explored the inner beauty of NFC reflected in factors such as closeness (inherent security), simplicity (zero configuration and zero administration), high value but humble (tiny size with memory capacity), dual personality (active and passive behavior), intelligence (low power consumption), and communicativeness (talks to RFID equipments). Nonetheless, there is an unmatched, eye-catching aesthetic simplicity that can dramatically impact user perception and that has not been mentioned above: its natural language, which breaks all linguistic barriers. It is a language that consists of a simple hand movement, waving or touching, while holding an NFC-enabled device. This is attractive to people and has the potential to simplify their interactions with the environment: a person with an NFC-enabled device by simple wave or touch is able to establish an NFC connection, that is, to interact with different devices in their ambient area (i.e., home, office, city). These NFC transactions can be used for user identification, payments, ticketing, or picking up information, among others, as will be seen.

Breakthroughs in antennas, integrated circuits, and batteries have led to the emergence of a new generation of NFC technology. This is the second aesthetic advantage: the technology being small enough to be embedded into common everyday objects, particularly mobile phones. This issue has also generated much interest in consumers. Thus, the catchphrase "all you need in one device" and mobility can be fulfilled.

The Growth of NFC

Beyond the nature of this technology, some organizations and initiatives are contributing to paving the way for its rollout in full swing. In 2002, Europe's Ecma International adopted NFC as a standard, and later in 2003, NFC became an ISO/IEC standard. Afterward, in 2004 the two main developers, Philips (NXP Semiconductors was founded by Philips) and Sony, together with Nokia founded the NFC Forum as a nonprofit industry association to promote NFC by developing specifications, ensuring interoperability among devices and services, and facilitating the NFC market (Forum 2004). These days, the

NFC Forum consists of 140 members that range from manufacturers to financial services institutions and application developers.

In 2007, the GSM Association (GSMA) started taking more interest in this standard under the initiative *Pay-Buy-Mobile* (GSMA 2007, Pay 2007). The main objective of this initiative is to enable the worldwide use of mobile phones for fast and secure payments. Thirty-four of the world's largest mobile network operators have worked together on this initiative. They are focused on defining a common global approach to overcoming the problems of expanding and bringing to maturity the use of NFC mobile phones to make payments.

The first NFC steps have come in the area of NFC payments, and apart from the GSM Association, other industry associations have wanted to accelerate widespread NFC adoption and usage for payments: the EMVCo and the SmartCard Alliance. The EMVCo LLC was launched in 1999 by the main card issuers (i.e., American Express, JCB, MasterCard, and Visa) to maintain and enhance the EMV™ integrated circuit card specifications. EMV™ is a global standard for chip-based credit and debit payment cards. The participants pursued to ensure interoperability for payment systems, including not only chip-based payment cards but also acceptance devices (Point-of-Sales terminals and ATMs) (EMVCo 1999).

Complementary to EMVCo's goal, the SmartCard Alliance, a multi-industry association with over 170 members worldwide, including participants from financial, government, enterprise, transportation, mobile telecommunications, healthcare, and retail industries, seeks to address NFC opportunities and challenges in industry (SmartCard 1997). In 2010, EMVCo, the SmartCard Alliance, the GSM Association, and the NFC Forum showed its first signs of collaboration by signing a memorandum of understanding describing their agreement to collaborate in further developments of the NFC market and NFC-based solutions. "Union is strength" describes the spirit of these associations.

NFC Embedded in Mobile Phones

A mobile phone seems to be the most valuable player or the best-quality signing instrument in the field of NFC applications. NFC technology is currently mainly focused on its use by means of mobile phones, a device that already travels with users and does not need any special card or equipment to make use of NFC services. This is of particular importance for users and environmental considerations due to the ever-increasing number of plastic cards, keys, and tickets, although there are additional reasons for using NFC-enabled mobile phones.

Why Use Mobile Phones

The following are a number of reasons that are highly significant for the usage of mobile phones in NFC services:

- **Anywhere personal device**: Recently, especially during the last decade, mobile phones have become very popular the world over, and almost everybody carries a phone everywhere and anytime. Hence, it is always with us, and furthermore, the number of mobile phones in use is rising every day worldwide. It meets the requirements of an all-in-one personal device.
- **Processing power and memory**: Mobile phones are increasingly being used for more than making calls; they are becoming little computers with their own operating systems and applications. A key feature of NFC is that users can also download software to them. In addition, other hardware elements, such as radios, video cameras, and GPS (Global Positioning System), among others, are increasingly being integrated on them, which can also be useful.
- **Network advances**: The mobile network is essential to enable remote application provisioning for NFC services. In that way, the continuous availability of the connected network also plays an important role. The use of advanced network technologies, such as 3G and 4G, makes it part of the NFC infrastructure. These technologies offer, on the one hand, greater security and faster speed than their predecessors, and on the other hand, they offer new applications: mobile TV, mobile Internet, multimedia, and so on.
- **User interface elements**: The user–phone interface elements of mobile phones—displays, touch screens, keyboards, sounds, and vibrations—together with features such as music, video, and photo player and recorder add significant and greater value to NFC services and applications. They provide a way of interacting with the NFC ecosystem.

- **UICC features**: NFC can leverage many features of UICC (Universal Integrated Circuit Card): integrity management of personal data, processing power, I/O circuits, memory storage, and extended security management functionality (Pay 2007). Traditional security features of UICCs are keys and algorithms for subscriber authentication; since cards are removable, they hold user subscription identity when changing phones, the access can be protected by the use of a PIN code, and it requires a PUK code for locking it out.
- Beyond that, advances in security management have also emerged for hosting applications provided by multiple third parties. For example, the UICC can be partitioned into several security domains, and each party's application can be run in each partition, maintaining data confidentiality and access protection among the parties' applications. The UICC can be seen as the secure element for NFC services. Furthermore, it is worth mentioning that UICC universal deployment and standards are expected to collaborate on global interoperability.
- **Mobile services**: NFC can leverage multiple services already available for mobile phones: request of ringtones, television, mobile Internet, media player, applets on SIMs to access Pay-TV content, Digital Rights Management (DRM) for storing rights for pictures, videos, sounds, and so on, and Over-The-Air (OTA) administration, such as activation/deactivation or personalization of the individual trust sectors via a trusted third party.
- **NFC already available in mobile phones**: Most manufacturers are already supporting NFC. Although it is still in its infancy, NFC has attained a certain maturity, and it is contributing to the creation of new applications, and thus a new business landscape. A first sign of maturity is the fact that these days many mobile phones equipped with NFC are already available on the market from most of manufacturers (for instance, BenQ, LG, Motorola, Nokia, Sagem, Samsung, and Sony Ericsson; Apple is already testing an NFC-enabled iPhone and so on).

As a result, NFC technology together with mobile telephony appears to have several advantages over traditional contactless technology, including high-speed peer-to-peer communications; media, Internet, and OTA services (used to configure your mobile phone with your applications) and access to facilities on the mobile phone (i.e., screen, keyboard, memory, processing power), among other attributes.

NFC Mobile Phones

An NFC-enabled phone is made up of the following components: a baseband with the handset operating system, an NFC unit/chip, an antenna to capture contactless data, and an NFC controller that interfaces with the UICC and handset memory. The NFC controller handles three modes of operation or communication: Card emulation, Read/Write, and Peer-to-Peer (P2P). These modes will be briefly described in the next section. The UICC element processes data coming from the phone and from the NFC chip, that is, from other NFC elements.

In 2007 the Nokia 6131 was the first NFC mobile phone to appear. At that time, some trials were conducted that highlighted some performance issues to be solved. Research was carried out to address higher security to transmit user's payments to banks or other type of information, and now these NFC handsets seem to be mature.

Other NFC-Enabled Elements

NFC was born with a silver spoon in its mouth; nevertheless, the NFC ecosystem needs to become widespread. The term *NFC ecosystem* has often been used in the literature to refer to an environment required for NFC services to emerge. NFC mobile phones present significant business opportunities for applications such as payments, ticketing, identification, physical access control, loyalty, and many other services. To create this NFC ecosystem, several "organisms" are needed, for example, mobile carriers, financial institutions, and mobile network operators. In addition, a habitat or environmental area inhabited by these organisms and human beings needs to be introduced. This habitat will be formed with NFC technology embedded in

- **Personal computers**: Including desktops, notebooks, netbooks, ultramobile PCs, tablets, or pocket PCs. The integration of NFC technology will enable future NFC applications such as the automatic synchronization of your phone calendar with your PC calendar.

- **Transportation**: NFC elements in means of transport such as buses, trains, trams, underground, taxis, airplanes, or public bikes. Also, NFC should be integrated in stations and airports.
- **Merchant and other service providers**: For example, shops, restaurants, hotels, drinks, and food packaging or dispensers.
- **Access control points**: Monitoring access to restricted areas and thus increasing security.
- **Consumer electronic devices**: For example, audio and video players, radio, or e-book readers. This will enable easy delivery of audio and video data or documents.
- **Urban settings**: Posters, information tags located in public places and elevators to provide new existing services such as for entertainment events and tourism.

Three Modes of Operation

This section describes the three modes of operation of an NFC-enabled device (Tags 2009). This classification helps in understanding how the wide range of applications works. These three operation modes have emerged to define the main use cases for NFC: card/tag emulation, reader/writer mode, and P2P mode.

Mostly, NFC-enabled devices can operate in reader/writer and P2P mode, and may function as contactless cards (card emulation mode); whereas NFC tags are passive tags that store data to be read by an NFC-enabled device.

Card/Tag Emulation Mode

In this operation mode, an NFC mobile phone can emulate a contactless card, for example, to purchase goods and services, to access services in public places in public transport. Thus, NFC mobiles phones can communicate with merchant Point-Of-Sales (POS), ticket machines, or any other object. To date, this is the most widely adopted mode, and it can leverage existing RFID-based equipment. It is worthy of mention that today contactless cards are already being used in payments and ticketing.

This mode, although it behaves like traditional contactless cards, can go beyond the functions of the traditional contactless cards:

identification, authentication, and data storage. This can be mainly achieved by means of NFC-enabled mobile phones. For example, the read information can be displayed on the mobile phone screen, and user authentication can be requested.

Reader/Writer Mode

This operation mode allows applications to transfer data in a nonsecure way. The NFC-enabled device operates actively, and it can read or write passive NFC or RFID tags. These tags could be embedded in a smart poster or any other everyday object. Mostly, the application data information stored on them can be the following: a URI, that is, a string of characters used to identify a name or a resource on the Internet, such as an URL, a telephone number, an SMS, or an e-mail address; general text; business card information, for example, virtual business cards, called vCards, with personal details; handover parameters for Bluetooth or Wi-Fi; and a signature, or a combination of them.

Peer-to-Peer Mode

The third mode of operation is the peer-to-peer (P2P) mode, which supports local bidirectional communication between two NFC devices, device link-level communication. Two or more NFC devices can exchange information: vCards, digital photos, sharing Bluetooth or Wi-Fi set up parameters. This is not supported by the contactless communication API.

Operation Modes: NFC versus Smart Cards This section explores the advantages of NFC compared to contactless smart cards in terms of operation. The following advantages can be highlighted (Essentials 2008):

- **Interactivity**: Thanks to phone interfaces and features, the card emulation mode can be used interactively by the user. For example, services can be presented visually on a mobile phone, or users can activate or deactivate an NFC application through the phone interface (i.e., touch screen, keyboard).
- **Management of multiple contactless applications**: This management is already supported by contactless cards, but

the continuous mobile network capabilities add functionality to this multiapplication management, mainly functionality related to security and risks with NFC services. Service providers can provide a trusted execution environment, assignment of trusted areas, application downloads, user personalization, and service locking/unlocking. From a user perspective, users can ask the service provider to stop NFC applications when a terminal is stolen or lost; that is, the UICC data can be modified over the air.

- **Management of user preferences**: The continuous communication of mobile networks offers new management tools from a user and service provider viewpoint. Users can access their personal data in real time and can change the information they would like to receive. On the other hand, service providers, with user consent, can retrieve information about NFC service usage records and can send users customized information.
- **Coexistence of NFC services but provided by different third parties**: Compared to a contactless card issued by a single service provider, an NFC mobile phone can manage and operate services from different service providers. This can be achieved because the UICC can be divided into security domains.

NFC has evolved to the point at which it offers significant benefits in a wide range of applications. In the following text, some applications are introduced to provide a glimpse of the goals and capabilities of NFC technology.

NFC Applications

ATMs revolutionized the way people got money, and yet they were met with great initial skepticism. In the past, "getting money from the wall" was considered to be ridiculous, whereas today we cannot think of living without them. Likewise, credit or debit cards also became a milestone, and initially the skepticism was fueled by the fact that people could not think of paying something with a simple piece of plastic. NFC is intended to provide the next generation of "gadgets" for paying, among other applications. Payment, ticketing, and loyalty applications are perhaps the reference applications of NFC technology, but

Projects:
● Big number
○ Medium number
◉ Small number

Figure 3.1   NFC trials and pilot projects in the world.

in this section we will see that NFC is much more than a payment or access technology.

In October 2008, a web-based publication called *Near Field Communications World* reported more than 100 NFC trials and projects in 38 countries. Figure 3.1 shows the distribution of NFC trials and pilot projects providing different NFC services in the world. As can be observed, Europe and Asia are taking the lead in NFC, whereas North America, the Middle East, and Australia are taking a backseat.

Mobile Payments/E-Wallet

Oddly enough, although the first plastic cards appeared in the United States in 1950, the breakthrough was achieved with the introduction of phone smart cards in France by 1986 (Finkenzeller 2003). Now, mobile phones can replace these cards.

How NFC Mobile Payments Work   In general, a mobile payment refers to the use of a mobile phone to perform a payment transaction. The transaction starts by waving the phone close to the POS reader. This transaction generally involves submitting a user's credit card information securely to a bank provider. Hence, upon acceptance, the bank will transfer the user's money to the corresponding person or institution. This transaction should be completely secure. The UICC can be seen as the secure element for NFC services and later, we will see other participants providing security in this mobile payments arena.

Essentially, translating a mobile payment into NFC language, a bank payment application should be installed in the UICC of an NFC mobile handset. By means of the Single Wire Protocol (SWP), standardized by ETSI/3GPP, the UICC-based payment application and NFC secure element within the handset are communicated. Hence, this secure element will be responsible for handling the payment from the end-user mobile handset to the merchant terminal or POS. Once the customer's bank and personal information is securely transferred to the POS, several data communications are sent to other external entities to securely support payment transaction. This functionality is rather similar to the current functionality, although more participants emerge.

Infrastructure Requirements   The mobile NFC ecosystem entities are shown in Figure 3.2. A key new entity in this ecosystem is the Trusted Service Manager (TSM), which securely distributes and manages services offered by service providers (i.e., in banking, transport, loyalty) to the mobile network operator's (MNO's) customers. However, not all NFC services should be handled by a TSM; there are some services that can be offered directly to the users without any risk (i.e., getting information from a smart poster). This ecosystem is valid for all NFC applications.

Traditional credit or debit card payments include customers, merchants, and financial institutions such as banks and payment solution

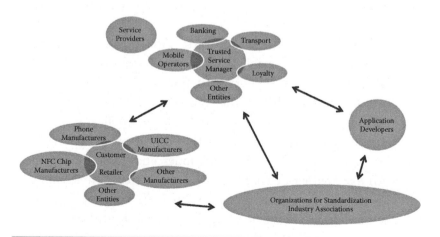

Figure 3.2   Entities involved in the mobile NFC ecosystem.

companies. In mobile NFC payments, new participants appear: MNOs; UICC, NFC chip, readers, and phone manufacturers; and TSMs. The following requirements have been envisioned in the corresponding entities (Pay 2007; Mobile 2007):

- **Payment Networks**: Companies such as American Express, JBC, MasterCard, Visa, China UnionPay, and so on can maintain their payment networks, although perhaps it is foreseen that they can provide other services to their banks, for instance, approval, certification, or definition of new requirements for different payment modes. These companies provide the payment platform for CIBs (Card-Issuing Banks).
- **CIBs**: They are service providers that need to provide NFC payment services to merchants and customers. They will be responsible for the issue of the payment application and for establishing a formal agreement with customers (similarly to the traditional cards' agreements) and merchants.
- **Merchants**: As the provider of the goods and services being purchased by the customer, they need to be equipped with NFC-enabled POS terminals. The RFID-based POSs are also valid. Similar to today's operation, the merchant has to sign a contract with a service provider that offers a merchant account and credit or debit card processing services. The functions of service providers do not change, but they do for customers.
- **Customers**: They need to sign a service agreement with a CIB so that they can have the corresponding application installed in their NFC mobile phones.
- **Mobile Handset Manufacturers**: Apart from providing standardized NFC-enabled mobile devices incorporating the UICC, they should support trusted execution environments, software download and management of multiple trusted applications, and multi-issuer coexistence applications. The term *standardized devices* means that they should support the corresponding NFC operation modes and communications with existing legacy contactless infrastructures. In addition, they should support interaction with NFC services by using their user interface functionality. These mobile phones will be provided by an MNO.

- **MNOs:** They must provide and maintain a network infrastructure to secure OTA deliveries and maintenance of the payment application on the UICC and provision of its security domain. Furthermore, they must provide mobile services to customers as well as NFC handsets with the required UICC smart card. Mobile payment services are to be securely provided and maintained by the TSM.
- **TSM:** The TSM must be in charge of securely distributing and managing services offered by service providers to the MNOs' customers. It should provide a mechanism to enable secure NFC services. In the past, mobile networks and services were regulated, and mainly locally controlled, over proprietary network platforms; today, the reality is that these new services are producing significant vulnerabilities, and interoperability needs to be addressed in some cases. As a result, the GSM association introduced the concept of TSM in 2007. TSMs are new entities in the NFC ecosystem, and they play a key role in payments, acting as a "link" between CIBs and MNOs. On the other hand, payment services regulations have been proposed to be managed by TSMs. Any above-mentioned entity as well as the government could be interested in playing this role, and to date it is not clear who will manage these responsibilities.

Mobile Payment Specifications   In an effort to achieve interoperability, all entities in the ecosystem must agree to operate under uniform technical operating and security specifications. NFC is already an Ecma and ISO/IEC standard, and it is compliant with ISO/IEC 14443. The NFC Forum has defined common data formats to ensure interoperability between devices and services (Forum 2004). In addition to these standards, several associations and initiatives have been working on the standard definition for secure payment transactions.

The ETSI's Smart Card Platform (European Telecommunications Standards Institute) deals with the UICC specifications and the connection between the NFC chip and the UICC (ETSI 2000). The specifications of security domains and required application programming interfaces to run on the UICC have been defined as Global Platform's specifications (Global 1999). On the other hand, the EMVCo payment standards body, owned by Visa, MasterCard, American Express, and

JCB, is responsible for developing and maintaining the EMV™ payment specifications. All these standard bodies together with the GSM Association are collaboratively working on worldwide payment systems.

The GSM Association represents the interests of the worldwide mobile communications industry with more than 700 members over 218 countries and serving more than 2 billion customers (82% of world's mobile phone users). In 2007, under the Pay-Buy-Mobile initiative, they defined a global approach to enable fast and secure payments using NFC technology that was supported by financial institutions and leading mobile manufacturers. One of the main objectives of this association is to lead standardization activities to define a common set of requirements and specifications for payments worldwide. We believe that this initiative sets the pace for mobile payments to become a reality: a new business model for mobile and financial industries.

User Trials and Pilot Projects   The first examples of NFC in use have been trials for payment and transport ticketing in 2005. The trials related to paying for transport tickets are reviewed in the next section; here we address the trials for contactless payments other than those. Since 2005, many trials and pilots have been performed that show how many countries are moving toward NFC adoption. Some of them even have been followed by commercial deployment.

In 2009, there were one or more user trials in at least 38 countries in Europe, the United States, Canada, Asia, and the Middle East. In particular, most of them have been focused on payments. Numerous trials are being carried out, and many others have been completed with successful outcomes. The very early trials were performed by using NFC stickers attached to mobile phones; currently, most of them are under way with real NFC phones.

It would be impossible to cite all the NFC trials or pilot projects in the world. As a result, a few trials have been chosen due to their pioneering efforts, technologically relevant initiatives, or some other reason.

Japan's Mobile FeliCa—Osaifu Keitai   Although FeliCa is an RFID-based contactless smart card technology, NFC is compatible with this technology, and FeliCa's success dates back to more than a year ahead of NFC (Felica 2010). Contactless payments are not new, and many solutions are being used all around the world based on RFID. Japan is the leading country in adopting these contactless solutions. In 1995, Sony completed the development of FeliCa technology, and it started being used in the Octopus contactless card in public transportation. Octopus was the first contactless smart card in the world. It was rechargeable, and initially it was used to pay transport fares. Later, in 1999, different trial uses of the FeliCa card as an employee ID and "e-wallet" came out. In 2002, a joint venture between Sony, NTT DoCoMo, all Nippon airways, and more than 50 companies launched Edy, a prepaid electronic money service using FeliCa technology to pay in shops, restaurants, taxis, and in many other places.

Earlier, in 2001, Sony and JR East railway network worked together to deploy contactless ticketing, and they launched the Suica (Super Urban Intelligent Card) contactless card to be used in Tokyo's railway network. It was an interoperable card to be used not only for transport ticketing but also to pay in affiliated stores in railway stations. Suica launched the market for electronic money in Japan. After the success of Edy and Suica, a big effort was made to integrate FeliCa technology into mobile phones between Sony, NTT DoCoMo, and JR East railway network. This effort involved not only technology advances but also the identification of a business model for all parties; for example, NTT DoCoMo and Sony established a joint venture to license and manage the use of the mobile FeliCa technology. This venture was called FeliCa Networks, and it functioned as the Trusted Service Manager (TSM). This TSM was responsible for computing revenues from companies that paid license fees, from providing services every time a user downloaded an electronic wallet application to providing other services. Thus, the companies were able to split profits between them.

The first FeliCa mobile phones were launched in July 2004 with a mobile version of Edy preinstalled as an application. This initiative was called "Osaifu keitai," meaning Mobile Wallet. The infrastructure for enabling payments was already there. In Japan, the success of this technology has been described as a key and expensive strategy undertaken by Sony and NTT DoCoMo. They made sure that retailers and POSs had FeliCa's technology available ready to read, initially FeliCa-based mobile cards, and later mobile phones. This met with rapid adoption. Visa Wave and MasterCard Paypass were also involved in this adoption. Afterward, credit services were provided to allow for mobile credit card transactions with and without authentication.

France's Payez Mobile   Payez Mobile is a field trial launched in November 2007 by six large banks and four mobile operators in France. It involved a thousand participants and 200 stores in the towns of Caen and Strasbourg. In this trial, participants were provided with mobile phones from Motorola, Sagem, and LG together with an NFC chip from Inside Contactless and Gemalto. On the other hand, Oberthur Card systems provided the SIM/UICC cards and the OTA application management platform. The starting point was that the bank payment applications—MasterCard PayPass and Visa VSDC—that were already installed in the UICC, although users could download other applications such as for ticketing, loyalty, or accessing. This trial allowed supporting customers handling multiple applications in a mobile phone.

The results of the experiment were positive. Thirty percent of the participants regularly used "Payez Mobile," and around three transactions per user were carried out after some months of experience. Seventy-five percent of these transactions were with amounts under 20€. The main feature of this project is that for the first time ever, many banks and operators were brought together. (Project page: http://www.payezmobile.com/)

France's Cityzi   In mid-2010, a first precommercial NFC pilot took off in Nice (France), involving a new brand called Cityzi. The objective was to discover consumers' behavior. Several trials have been placed in France during the last years, but at this time NFC phones were on sale to the general public for the first time. The name Cityzi was used to indicate the availability of standard-compliant services, in posters and billboards. Users could make payments, purchase transport tickets, and access real-time travel information at each bus and tram departure point in Nice, thanks to 1,500 NFC and 2D barcode-enabled information points across the local transport network. This information included services provided by the city council, events lighting, and the latest news articles from the region's major daily newspaper. In addition to that, consumers could collect loyalty points upon purchases. (Project page: http://www.cityzi.orange.fr/)

London's O2 Wallet or "Pay & Go"   Between 2007 and 2008, a six-month trial was undertaken using a Barclaycard payment application in the O2 Wallet. Users could make payments for £10 or under at retailers in and around London, without any PIN or password for confirming payments. This trial involved 500 participants. Initially, they were credited with £200 in their mobile phones, and they could check the available funds. The next steps would be introducing a password or PIN for making payments, allowing purchases over £10, and having credit funds capability.

Belgium's PingPing   Belgacom, a Belgian telecom operator, tested a series of mobile payment initiatives with a number of high-profile partners, including Accor Services, Coca-Cola, and supermarket chain Delhaize. A new payment's logo and brand called PingPing was created to indicate to customers where to make mobile payments. A thousand employees from Belgacom and neighboring companies used their mobile phones (an NFC tag was attached to the back) to purchase meals, refreshments, and shopping in shops, supermarkets, and restaurants that accepted contactless payments. For example, Coca-Cola added NFC readers to drinks-dispensing machines, and Delhaize also added readers to checkouts near Belgacom's premises. Tunz, an electronic money issuer that holds a European e-money license, supported NFC payments. (Project page: http://www.pingping.be/wp/about/)

Spain's Mobile Shopping   In May 2010, a six-month trial for NFC payments in 500 shops was performed in a town near Barcelona, Spain. About 1,500 people participated in this trial. The trial was supported by Telefónica, a Spanish's bank (La Caixa), Visa, Samsung, Gemalto, and NXP Semiconductors. The mobile phone model used was developed by Samsung, the secure element was on the UICC developed by Gemalto, and NXP Semiconductors was in charge of the NFC phone chip. The TSM role was played by Telefónica, and the infrastructure of Visa payWave was used. (Project page: http://www.mobileshopping.es/)

Canada's Mobile PayPass™   In 2009, Citi Cards, MasterCard, and Bell Mobility carried out a trial of Mobile PayPass™ with 75 employees in Canada to make payments of just under a certain amount of money (around US $20). These transactions were recorded for further analysis. This four-month trial was very successful since participants proved that the payments were easy, fast, and secure.

In the same year, another trial was carried out by the mobile operator, Rogers Wireless. This six-month trial tested the OTA provisioning and NFC payments. About 250 users could pay by using a Motorola prototype (Inside Contactless developed the NFC chip) in 70 merchant locations by using Visa payWave. In this case, Gemalto, a leader in digital security, was the TSM, and the Royal Bank of Canada also participated. Trial participants conducted more than 2,100 retail-payment transactions during the six-month trial. A post-trial survey placed user satisfaction rates at more than 70%, according to the bank. Theirs was one of the first NFC initiatives in Canada.

America's "TAP & GO" In 2007, for 6 months, selected Citi MasterCard cardholders with Cingular Wireless accounts participated in a trial for "tap and go" payments through NFC Nokia mobile phones in New York. These were provided with MasterCard PayPass contactless payment capability. Participants could pay for purchases at any merchant accepting MasterCard PayPass and in the New York City Subway for fare payment. NXP Semiconductors developed the NFC chips for the NFC-enabled mobile phones. Giesecke & Devrient developed the OTA personalization software that allows for the process of initializing your mobile phone with the information it needs to conduct secure transactions. The contactless payment transaction is automatically charged to the user through the same secure MasterCard payment network that processes traditional credit card transactions.

In Dallas, the first implementation of Giesecke & Devrient's secure chip management platform—the OTA platform—was tested with Nokia mobile phones, MasterCard PayPass, and Bank of America. The wireless service and network access was provided by 7-Eleven through its prepaid Speak Out wireless program in 2007. Therefore, users could use mobile phones not only for payments but also for personalizing them with different applications. Up to 500 participants were provided with NFC Nokia mobile phones to make purchases. Initially, they had to make a one-time request to their banks to register for the payment service. Data was sent by the secure chip management platform over 7-Eleven's network. Hence, the PayPass payment application (with password protection) was installed inside the mobile's secure element. The trial was run for six months, and users could make purchases at around 32,000 locations that accepted

MasterCard PayPass. This trial showed that consumers were still not very familiar with downloading applications to mobile phones, and many issues needed to be improved: the OTA platform for download-ing applications, easier ways of downloading an application, and not using text-based but a simple call to request an application.

China In 2008, a payment trial was performed in Shanghai. Eight hundred participants could pay at around 300 point-of-sale terminals (i.e., restaurants, shops, ...). This initiative was by China Union Pay, NXP Semiconductors (NFC phone chip), Bank of China, Industrial and Commercial Bank of China, Bank of Communications, and Industrial Bank and Shenzhen develop-ment bank, among others. The phone used was Nokia 6131. There were several bank applications inside the secure element and users tested the OTA service.

Mobile Payments in Commercial Operations After user trials, some countries have already adopted commercial deployments around the world. The world's first NFC commercial service was in Malaysia in 2009 with Maxis (mobile operator), Maybank, Visa, Touch'n'Go, and Nokia. Users could pay at shops, restaurants, public transport, highway toll gates, and car park facilities.

In Japan, they are mainly using FeliCa technology. Some other examples of commercial deployments are in France, Czeck Republic, Turkey, Beijing (China), and so on. In July 2010, Avea and Garanti Bank plan to start a commercial rollout to support payments in more than 15,000 POSs with MasterCard Pay Pass and more than 10,000 with Visa PayWave in Turkey. Users can also earn discounts and gifts. It uses Gemalto's Upteq N-Flex device, a SIM together with an antenna compliant with the SWP that is integrated in the phone.

Conclusions Overall, a high percentage of the participants are inter-ested in using mobile phones for payments when it becomes com-mercially available. From 2005 up to now, different trials have been carried out to improve all the entities' tasks in the NFC ecosystem, to evaluate feasibility, and ease of use. The technical feasibility of NFC for payments and user acceptance has been positively evaluated.

It seems that the following risk management policy is being used: for low-value payments (that is, 20€ or even more) transactions are PIN-less, whereas payments exceeding a certain amount of money require a PIN.

There are several future applications regarding money. One may be of special interest: mobile remittance, that is, transferring money from your mobile phone to another device securely.

Mobile Ticketing/E-Ticketing

NFC offers new opportunities for many ticketing scenarios; paper tickets will be hardly used, since it will be possible to read your ticket data from your NFC phone.

Public Transport—User Trials

Hanau's Bus Ticketing   The first examples of NFC in transport ticketing came in 2005 on a local bus network in Hanau, Germany. This one-year trial was based on Nokia handsets provided with an NFC shell for counting the check-in and check-out in buses. The Rhein-Main-Verkehrsverbund (RMV), a regional public transport authority, tested the NCF mobile phones for electronic ticketing with 146 customers on 15 buses. The participants only needed to touch the mobile phones to the NFC readers (already available for contactless cards) as they got on and off the bus. This trial was supported by RMV, GmbH, Hanauer Straßenbahnbetriebe (HSB), Nokia, Philips, and T-Systems. Nokia provided the ticketing application stored in the integrated smart card controller in the mobile phone that enabled storage and management of tickets. The next step would be to pay for bus fares. The technical feasibility of NFC and user acceptance was positively evaluated.

London's Tube, Buses, and Trams   In 2008, a six-month trial was run in London's tube, buses, and trams in partnership with O2, Barclaycard, Nokia, Transys, Visa, and AEG using 500 mobile phones with built-in Oyster card technology. Each NFC phone was provided with the "O2 wallet" application for payments, public transport, event ticketing, and smart posters. In case of a phone call or text message, users were able to answer the call and continue with a transaction.

Sports Ticketing   One of the first trials related to sports ticketing was in Kerkrade, the Netherlands, in 2005, and it lasted about 6 months. The mobile operator was KPN, and the service provider was Roda JC football club. Fifty users participated in this trial, and were provided with a Nokia phone. Bell ID was the implementer, and NXP Semiconductors developed the NFC phone chip and the secure element. These users had their season passes stored on the phones to enter the stadium. They could also use their phones as prepaid cards for use at food stands and shops in the stadium. They were able to check their balance.

Travel Ticketing   Apple seems to be interested in having travel ticketing at mobile phones, as well as boarding passes and check-in technology. This application has been called iTravel. iTravel is an NFC-based application, with its own interface to address travel matters as airline check-in, baggage identification, electronic ID, car rentals, hotel reservation, and so on. All you will need, for example, is an image taken with the iPhone camera of your boarding pass, and readers should be able to extract this information from the image by using an optical character recognition, a barcode reading software, and/or a QR code. (More information: http://www.patentlyapple.com)

Accessing Other Transport Information   A mobile phone can also be used as a resource for accessing transport information, such as timetables and changes in flight gates, among other services. In addition, it can be used for transmission of electronic travel documents to security groups in an airport, train, station, etc.

Keys, Personal Identification, and Easy Login

Apart from the most popular applications and following the steps of contactless cards, this is one of the main fields of application for NFC technology. NFC offers great potential in contactless identification and control access solutions.

Contactless cards are typically used in transport, and sometimes they are not used for buying a ticket but as a transport ID. Contactless keys have been around for many years, and they can also benefit from NFC technology. For example, your NFC mobile phone can be used for entering your home and your office or it can be used as the room key in a hotel.

Several trials between 2005 and 2006 were carried out in the city of Caen (France), not only for payments but also for car parking access and getting tourist and transit information via smart posters. The participants were Philips, France Telecom, Orange, Samsung, Groupe LaSer, and Vinci Park (retailers). Users were able to make payments in an underground car park; the town hall; a bus stop that could transmit timetable information; a cinema poster that downloaded video trailers to users' mobiles; a local supermarket, where people could pay for their groceries with a mobile phone; and in a tourist information point. Users could receive information about the place they were in by either an SMS or a phone call. Around 200 participants took part, and they only had to touch their mobile at the *Flytag* logo at these locations. In this trial, the OTA download of applications on a GSM network was available, for example, to install the application "car parking access." Once the NFC connection was made, the appropriate application was launched. Moreover, payments were supported with levels of security.

Due to NFC's simplicity of use, by waving or touching an object, an NFC mobile phone could be used for logging into a computer.

Tourism/Events/Information

In peer-to-peer mode, an NFC mobile phone can be used for accessing information. For instance, it can be used as a resource for accessing transport information, such as timetables, changes in flight gates, and so forth.

A smart poster is a poster with embedded RFID-based tags. With the read-write mode, an NFC device can read an RFID tag in a smart poster that allows users to download tourist information, concert information, a URL for a movie trailer, and so on. Even in museums, tag information related to a painting or any other piece of art could be captured and displayed in a mobile phone.

A smart coupon can be an NFC tag that can be embedded in a magazine, in a food item, in a smart poster, or in a public place, so that with your mobile phone you can get product information or discounts. Another trend, which is very popular, is the smart storefronts that allow merchants to attract new customers. Users can get information on restaurants around, call the restaurant, or view a map of its location.

Healthcare

NFC can also play an important role for healthcare at home, in hospitals, or in any other point of care.

NFC Opportunities in Healthcare   NFC also offers big opportunities in healthcare solutions:

- **Identification and health monitoring**: Services for automatically transmitting patient data securely over the air can play an important role, in particular, with an aging population all over the world and also for chronically ill patients. For example, users could send their blood pressure to the doctor by simple waving an NFC phone. This is an example that could pave the way toward telemedicine. Some products such as the UA-767 blood pressure monitor, a device for telemedicine, could soon be available in the market provided with NFC communication and new NFC services.
- **Medical record storage**: Health-related information could be stored in an NFC device to be rapidly evaluated at point-of-care centers, hospitals, or in an urgent situation. For example, data such as blood type, diabetes, asthma, and allergy conditions, among others, could be stored. Moreover, health-related parameters measured at homes could be stored over time and once in a point-of-care center, these data could be processed by the doctor to evaluate medical therapy or diagnosis.
- **Better diagnosis**: Longer and more complete medical reports can be a great help in providing better diagnosis.
- **Inventory tracking**: Some medical machines require cartridges; NFC could take care of the number available.

- **Medication care**: NFC could identify patients and their medication, and it could be used to dispense medication automatically and securely. NFC can also provide safer medication care at hospitals and at home, for example, for blind people to provide them with audio medicine identification through NFC.
- **Blood transfusions**: NFC could be used for safer blood transfusions.
- **Getting or storing other health-related information**: For example, NFC could be used for storing doctor appointments. Doctors or any medical practitioner could also take photos of a health problem in skin with their mobile phones, for example, to be uploaded automatically to a user's file.
- **Tracking medical practitioners' location**: Doctors or any medical practitioner could touch an NFC reader in a hospital floor or room to keep track of their positions or to know exactly which patient is being treated.

User Trials In 2009, a healthcare application was tested in the United Kingdom. The system enabled home health care workers to download patient records and care requirements by swiping an NFC mobile phone over NFC tags installed in a patient's home. Patients could also use the system to bring up details for their caregiver's next visit or their next medical appointment, to request an appointment or call back by choosing from a drop-down menu on the phone and to alert someone in case of emergency. Also, these services could be used by simply waving the mobile near a certain NFC tag. This test was developed by 02, a mobile operator, in collaboration with Reslink, a mobile phone solutions provider.

Several Applications

The purpose of this section is to describe a group of applications, not included earlier, in which NFC technology can also be used. These applications can be at different stages of advancement in different countries, research initiatives, trials, or just brainstorming ideas that could be future applications.

Loyalty Programs Together with the payment and ticketing methods, this is one of the most widespread applications that NFC can offer. Loyalty points can be stored in your mobile phone to get benefits from retailers. It is worth mentioning the key role of this technology contributing to a friendly environment by reducing the number of plastic cards.

Easing Parking A French group from the school Polytech'Marseille has designed an application to ease the pain of parking in busy urban areas. This application provides a package of services to help drivers: quickly find a parking lot, store a virtual parking ticket in their mobile phone, localize and retrieve the location of their car in the parking lot, track their parking bill in real time, and use their mobile phone to pay for the parking service. In addition, users can also collect special offers in the car park to be used at nearby shops and restaurants.

User-Centric Information in an Elevator In public places such as shopping centers, buildings, or elevators, general information is usually displayed through screens. With NFC technology, it is possible to adapt such information according to the user characteristics, needs, and interests and also according to context or ambient conditions. In that way, NFC opens up real possibilities of developing new user-centric services that could change the relationship, for example, between sellers and buyers or business people and clients.

LCD displays are increasingly being installed in public elevators, mainly at airports, hotels, hospitals, and offices. Generally, they display general information of particular interest; however, with the use of NFC a new collection of services can be offered to users. This can be achieved by simple placing an NFC reader inside the elevator or an NFC-enabled display. In this way, personalized multimedia content access while moving up and down in the elevator could be provided in displays. According to the user profile and preferences this information could be adapted in both content and formats (i.e., for visually impaired people, children). The personalized multimedia content could range from news headlines or events to advertisements. These can be adapted according to user preferences: sports news, classical music, and weather forecast, among others. As per another service, knowing in advance the user profile in an elevator could help decide what emergency protocol should be used in case of an emergency situation, for instance, for an elderly or handicapped person.

In the Automation World   In the automotive industry, a swing toward NFC can occur. NFC can be integrated for door opening and many other in-car services. For example, the car can be unlocked as the driver approaches, or the trunk can be opened when the user waves his or her mobile phone near it; after user identification, the car can be started by only pressing a button, and the car could be immobilized if it was stolen. Of course, this involves some security techniques that should be taken into account, for example, locking all NFC services if the NFC mobile phone is stolen.

Multimedia/Entertainment/Gaming   A simple example can be NFC speakers playing music sent from NFC mobile phones. Another example appears when thinking of Bluetooth and the tedious pairing process. Many multimedia elements often come with Bluetooth technologies. In this case, NFC can be used for pairing the two Bluetooth devices easily.

NFC could enable easy delivery of audio and video data or documents from our phone to audio and video players or e-book readers. It would also be useful if game consoles were provided with NFC. In this way, through your NFC mobile phone, you could download your games (with the corresponding DRM) and pass them directly to your console.

Remote Control   An NFC phone can communicate with other NFC-enabled computing devices in a home and act as a remote control.

Configuring Wireless Protocols   Sometimes it is really difficult to configure wireless Bluetooth and Wi-Fi, among others. And users often must spend considerable time to configure wireless devices to work with these wireless networks. Users can put two devices near each other and let them exchange the required settings via NFC. NFC enables more simplicity by allowing them to connect to wireless networks without introducing network parameters.

Multipurpose Projects

Apart from the trials, pilot projects, and upcoming applications already seen, it is worth pointing out that several research projects have also been underway in Europe to promote the use of NFC and to shed some light on some research issues in NFC.

EU Project—SmartTouch

The SmartTouch project has been one of the largest efforts on piloting NFC technology in the European Union (2006–2008). Many different pilots were carried out to study the role of NFC services in diverse domains such as city life, home, well-being, and health. Twenty two partners from eight European countries participated in this project under the Eureka and Information Technology for European Advancement (ITEA2). The project involved industry, research institutions, and public organizations (i.e., Alcatel-Lucent, Gemalto, Nokia, Philips, Fagor, RMV, Telefónica, ToP Tunniste, RMV, VTT, Telvent, Leuvent, Oulun kaupunki, Ikerlan, Robotiker, Visual Tools, and so on). For example, during this project several pilots were conducted in the City of Oulu and other partners' countries. (Project page: http://ttuki.vtt.fi/smarttouch/)

EU Project—StoLPaN

StoLPaN (Store Logistics and Payment with NFC) is a project supported by the European Commission's IST program. Twenty-three partners from different European countries collaborated on the project, such as Motorola, SafePay Systems, Deloitte, Sun Microsystems, NXP, Ennova-Research, Sheffield Hallam University, etc.). This project's goal is to define open commercial and technical frameworks for NFC services on mobile devices. These frameworks will facilitate the deployment of NFC mobile applications that are mobile device and service independent. In addition, they will also provide mechanisms to deliver securely third-party applications into the UICC and to create a multiapplication environment in the UICC (Project page: http://www.stolpan.com/).

Benefits of NFC Systems

The use of NFC-enabled devices in public and nonpublic places provides a multitude of benefits to all those involved (that is, users, consumers, and companies). Finkenzeller pointed out several

benefits of RFID systems that can also be associated with NFC
(Finkenzeller 2003):

- **Benefits for passengers**: Cash is no longer necessary, if fares
  change, prepaid contactless smart cards remain valid; there
  is no need to know the precise fare, and monthly tickets can
  begin on the day of the first use.
- **Benefits for the driver**: The driver no longer sells tickets, so
  there is less distraction. No cash need be kept in the vehicle,
  and daily income calculation is eliminated.
- **Benefits for the transport company and transport association**: Reduction in operating and maintenance costs of ticket
  dispensers and tickets such as costs in printing tickets; it will
  be easier to change fares, easier to apply discounts, and so on.

We can extend this list of benefits. Other benefits are highlighted in
the following list; some of them are exclusive to NFC, and others are
shared between both technologies:

- **Flexibility**: NFC can operate in three modes: card emulation, read/write and peer-to-peer, being a writer and a reader
  at the same time.
- **Reduction in the number of devices we carry**: Since an NFC
  mobile phone can hold one or more credit or debit cards, loyalty cards, boarding passes, etc., this also eliminates the problem of losing a credit card or reduces the risk of cash theft.
- **Reduction in times**: For transactions such as activating/deacti-
  vating a service, buying tickets, when boarding, when checking
  in and checking out, and so forth, time needed is reduced.
- **Reduction in costs**: Cash-handling costs in mobile ticketing,
  access keys, business cards, operational, and maintenance costs.
- **Security**: NFC offers higher security than RFID, because it
  requires the close proximity of two devices, and as a result,
  signals are difficult to intercept. Thus, this gives the technology some inherent security.
- **Simplicity**: The user only needs to tap, wave, or touch an
  NFC device to activate a service. This property is also characteristic of contactless smart cards. There is no need to sign a
  receipt and keep it for guarantee.

- **Speed**: In public transportation, in cinemas, when paying in a shop, and so forth. This benefit is valid for NFC and RFID technologies.
- **Accessing new convenient services**: For example, in an elevator, better and easier healthcare management, configuring a wireless connection, and so on.
- **Borderless services**: When traveling abroad, users do not need to get cash from the visiting country or they do not need any special equipment to acquire tourist information from monuments, posters, etc.

Conclusions

We are witnessing the second wave of interest in contactless mobile payments and ticketing. The first wave came out with contactless smart cards, and now it is time for NFC mobile phones. NFC technology together with other research advances are attempting to make NFC contactless payments and ticketing a reality. Payments, ticketing, and loyalty applications are perhaps the reference applications of NFC technology, since they have been perceived to be in great demand by consumers. As a result, the number of pilot projects introducing NFC payments and ticketing has been significantly higher than others.

On the whole, the trials have helped identify problems, successes, usability assessments, and required improvements in technological solutions and business models. For example, a noticeable result has been that NFC mobile phones have proved their value and have matured over these years of pilots. Today, NFC mobile phones are increasingly becoming commercially available, provided by most of the phone manufacturers in the world. With these pilot trials, the entities in the NFC ecosystem have also gained experience and learned lessons for the commercial rollout of these NFC services. The next generation of contactless payments is already around us thanks to the advent of not only these NFC mobile phones but also the already existing contactless infrastructure, new defined roles (among network operators, banks, and trusted manager services), and continuous technological breakthroughs.

Nevertheless, we cannot typecast NFC as a method of payment; NFC is much more than a contactless payment technology. NFC can open up real possibilities of developing new user-centered services that could change the relationship between service providers and customers in many domains and between us and our environment. It will enable a range of applications and services: proximity services such as getting information by touching smart posters or sharing information between phones or any NFC devices, being the boarding pass in an airport, functioning as a remote controller or a key, advertising services, location-based services, and so forth. To achieve these services and many more we may not yet have even thought of, it should be highlighted that notebooks, TVs, other computing devices, and noncomputing devices should also become NFC enabled.

There is still a long way to go to leverage all NFC features in many domains. Milestones along this way are building security and privacy NFC infrastructures, DRM services, standardization processes, and the complex service environment that should handle thousands of service providers providing services to several hundred million users within a few years.

We believe in the long-term potential of NFC to make a significant difference in the quality of life and comfort for many people and for our environment. This technology can be woven itself into our environment according to Weiser's vision of Ubiquitous Computing, and it can be the key technology for the Internet of Things. On the other hand, on the lines of "one picture is worth a thousand words," we would like to think that "one mobile phone is worth a thousand services."

NFC Mobile phone = keys + ID cards + cash + credit cards + debit cards + health smart card + travel card +......

Web References

The information of the different trials and pilots was mainly taken from the following websites:

- http://www.nearfieldcommunicationsworld.com/
- http://www.nearfield.org
- http://www.nfcnews.com
- http://www.nfctimes.com
- http://www.contactlessnews.com

- http://www.smartcardalliance.org
- http://www.nfc-research.at
- http://www.mastercard.com/paypass
- http://www.nearfield.org/

References

EMVCo. 1999. EMVCo organization. Web Page: http://www.emvco.com/

Essentials. 2008. Essentials for successful NFC mobile ecosystems white paper. Retrieved November 20, 2010 from http://www.nfc-forum.org/resources/white_papers/

ETSI. 2000. European telecommunications standards institute. Web page: http://www.etsi.org/

Felica. 2010. Sony FeliCa. Web Page: http://www.sony.net/Products/felica/

Finkenzeller, Klaus. 2003. *RFID Handbook: Fundamentals and Applications in Contactless Smart Cards and Identification.* John Wiley & Sons, New York.

Forum. 2004. NFC Forum. Web Page: http://www.nfc-forum.org/home/

Global. 1999. Global Platform Web Page: http://www.globalplatform.org/

GSMA. 2007. GSM Association. Web Page: http://www.gsm.org/

Mifare. 2010. NXP Semiconductors Mifare—contactless identification and verification. Web page: http://www.nxp.com/

Mobile. 2007. Mobile NFC services paper. Version 1.0. Retrieved November 20, 2010 from http://www.gsmworld.com/documents/nfc_services_0207.pdf

Pay. 2007. Pay-buy-mobile business opportunity analysis white paper. Version 1.0. Retrieved November 20, 2010 from http://www.gsmworld.com/documents/gsma_nfc_tech_guide_vs1.pdf

SmartCard. 1997. SmartCard alliance. Web page: http://www.smartcardalliance.org/

Tags. 2009. NFC forum type tags. Retrieved November 20, 2010 from http://www.nfc-forum.org/resources/white_papers/NXP_BV_Type_Tags_White_Paper-Apr_09.pdf

4

NFC APPLICATION DESIGN GUIDELINES

FELIX KÖBLER, PHILIP KOENE, FLORIAN RESATSCH, JAN MARCO LEIMEISTER, AND HELMUT KRCMAR

Contents

Introduction

Near Field Communication (NFC) has the potential to directly impact vast numbers of mobile phone users worldwide, since NFC-equipped phones enable them to bridge physical and virtual worlds (Rukzio et al., 2006). NFC can be used to (1) initiate a service, for example, by opening a communication link for data transfer, (2) enable communication between two devices, (3) or top off existing services, such as ticketing and electronic payment infrastructures (NFCForum, 2007). The challenges for companies interested in developing and marketing NFC applications lie in (1) being able to justify all initial investments in NFC infrastructure, (2) dealing with the insecurity of users accepting the technology, (3) not knowing which technology future users will employ in their daily lives, and (4) recognizing that consumers need to experiment with NFC to get a better idea of what it has to offer.

In order to enable companies to respond to these challenges, this chapter introduces a set of guidelines to design future NFC applications with a high user acceptance in consumer markets. A set of nonfunctional requirements and design guidelines are derived from four case studies conducted by the authors (Koene et al., 2010; Resatsch, 2010). They are meant to outline the positive and negative factors related to user acceptance of NFC-based ubiquitous computing artifacts in four example domains. Furthermore, the guidelines are exemplarily applied to the development of the NFriendConnector application in a prototype study. The NFriendConnector prototype is a peer-to-peer NFC application that enables seamless integration of users' off-line social interactions with their online social networks using a mobile phone. It allows users to instantaneously initiate and establish Facebook social ties using their mobile phones without having to incur any additional search cost. We considered the nonfunctional requirements and design guidelines in the conceptual design and development of the NFriendConnector prototype. The requirements and guidelines were interrelated to design rationales applicable for the NFriendConnector application. The application of the design guidelines in the development process of the NFriendConnector prototype allowed for a very high user acceptance of the application, as evidenced in a medium-scale laboratory experiment, despite employing NFC technology that was novel to participants.

The remaining sections of the chapter are organized as follows: The section titled "Theoretical Background" introduces the theoretical background on user involvement during the conceptual design and development phases for prototype applications. In the section titled "Deriving Design Guidelines for NFC Applications," we derive NFC application design guidelines and nonfunctional requirements from four NFC application case studies. The section titled "Guideline-Evaluation with the NFriendConnector" outlines how these requirements and guidelines can be transformed into design rationales for an NFC application, exemplified by the NFriendConnector development process. We end with a discussion of limitations and future research on design guidelines for NFC-based applications.

Theoretical Background

Companies strive to identify market and consumer needs right from the start, and to lower the high failure rates of new products (Gourville, 2006). A common way of determining these needs is to involve customers/consumers in the development process (Gruner and Homburg, 2000). According to Nambisan (2002), customers can be involved in generating ideas for new products, co-creating these products, testing finished products, and providing end-user support.

User Involvement

The inclusion of customers—users in the case of application development—in product development negatively impacts time to market and development costs. There are several concepts to the inclusion of potential users in the application development process that help reduce development costs and time to market. Three of these important concepts are outlined in the following section: (1) prototyping, (2) participatory design, and (3) lead users.*

1. **Prototyping:** A first approach to the integration of users into the development process of information systems begins with evaluating small low-tech prototypes. Prototypes build a basis for user experience, especially in the forms of preprototypes or paratypes, as explained by Davis and Venkatesh (2004). This method fully complies with design research approaches where the artifact (Simon, 1980, 1996) can be the later prototype. Various prototypes make it possible to identify design constraints and the contextual grounding of the design earlier on in the development process (Muller, 2003). The existence of prototypes during the development process additionally allows users to experiment with the technology and the concept of prototype applications. Prototypes have been a common component of the software development methodology (Floyd, 1983, 1989). Prototyping is a way to incorporate communication and feedback in software development. The emphasis must not be placed on building, but on evaluating

---

* The following sections are taken and partially adapted from Resatsch (2010).

prototypes, because this makes it possible to define the functional selection and determines further use. Bleek et al. (2002) claims that *traditional prototyping* is only partially valid for the development of Internet-based applications because of the diversity and complexity of Internet projects (e.g., portals, virtual online communities, social online networking sites, etc.). To what extent this is also valid for NFC is not clear. To help future users fully understand the potential of NFC applications, the use of experimental aspects can help them in an active way (Buchenau and Suri, 2000). Buchenau and Suri (2000) describe the use of experience prototypes to evaluate how users experience new products. They re-create and study simulated experiences, but do not conduct these in real-life situations—a step that is necessary to fully understand the consequences of ubiquitous computing. For NFC applications and products accessing this new technology, a *prototyping culture* is needed, as the market is yet to be defined. It is then important to align the various described forms of prototypes with adequate evaluation methods.

2. **Participatory design:** Another popular approach to integrating users is *participatory design* (Muller, 1993; Muller et al., 1991). Participatory design is a set of practices, theories, and studies related to end users as participants in the development of software and hardware computer products and computer-based activities (Greenbaum and Kyng, 1992; Muller, 2003). It also involves collaboration between users and developers in the design process. The concept has unresolved issues, such as participation of a nonorganized workforce or the need for visual and hands-on techniques, which violates the requirements of universal availability (Muller, 2003). Visual and hands-on techniques are also difficult in the case of ubiquitous computing technologies because of its intrinsically complex technological nature. Another weakness of participatory design concerns formal evaluations and data. A study on measuring success would require that a product be implemented and marketed twice: once with participation, and once without (Muller, 2003). This is almost impossible because of the high costs and huge effort involved. According to Blomberg and

Henderson (1990), the process of participatory design should be about improving the quality of the user's work life, using collaborative development, and creating an iterative process. These tenets may also serve as guidelines for developing a specific ubiquitous computing process. Participatory design commonly uses low-tech prototypes to create new relationships between people and technology. The end user may reshape the low-tech materials in a *design-by-doing* process (Bødker et al., 1995; Muller, 2003) resulting in a range of benefits, including enhanced communication and understanding, enhanced incorporation of new ideas, better working relations, and practical applications with measured success.

3. **Lead users:** The extent to which customers can optimize the product development process significantly varies. Von Hippel proposed what are known as *lead users*, to provide some insight (von Hippel, 1986, 1988, 2001, 2005); he also introduced *user toolkits for innovation* with which users can actively create preliminary designs, simulate or construct prototypes, evaluate functions, and then improve the design of the underlying product (von Hippel, 2001). *Lead users* state specific needs months and years before the mass market jumps on the bandwagon. Solving these needs is very important to lead users; thus, they are intrinsically motivated to assist in new product development (Urban and von Hippel, 1988; von Hippel, 1986). Furthermore, lead users can fulfill an important function in the prelaunch phases of a product and fuel the diffusion process in the postlaunch phase as opinion leaders (Morrison et al., 2000; Urban and von Hippel, 1988). Integrating lead users within the new product development process calls for initial prototypes as a first step: Given that NFC is currently only at an early stage of its development, potential users/clients are unaware of the purported technological potential of this novel technology. NFC technology might satisfy the needs of lead users; however, in order to explain its potential, even tech-savvy lead users need something to experience. The complexity of building an NFC application can also not be done by lead users through the mere use of tool kits in the first step, applications need to be designed and implemented in order to determine benefits and failures.

Deriving Design Guidelines for NFC Applications

It is necessary for developers to determine the system requirements for ubiquitous computing applications at an early stage in the development in order to cope with the aforementioned challenges. Requirements in system development usually specify the functionality that should be implemented and the way the user interaction should be handled for this implemented functionality (Kotonya and Sommerville, 1998). However, for many types of information systems, it is impossible to distinguish specific requirements of the software functionality from the broader requirements of the system as a whole. This holds especially true for ubiquitous computing systems that range from small embedded systems to large-scale infrastructures, making it difficult to gain information on detailed software requirements at an early stage. In particular, the user-centered focus of ubiquitous computing requires a more detailed examination of system requirements than in classic command-and-control systems.

Usually, a requirements specification includes both functional and nonfunctional requirements. A functional requirement specifies a function that a system or system component must be able to perform (IEEE, 1990). Nonfunctional requirements are defined as the non-obvious features and functions that do not serve a particular purpose, but are instead quality attributes. The diversity of ubiquitous computing systems leads to many requirements (e.g., visibility, conceptual model, etc.) that do not specify a function or a use case that can later be implemented in the application; they are instead integral parts of a successful infrastructure and express qualitative attributes (e.g., *haptic feedback*) and nonobvious features (e.g., everyday task specifics). For a system development process, these requirements are a basis for all applications used during and refined throughout the process.

A set of nonfunctional requirements for specific ubiquitous computing infrastructures can be derived from a limited number of showcase applications and used as basic requirements in the early development of similar ubiquitous information systems. In order to generate such a set of nonfunctional requirements for ubiquitous computing environments enabled by an NFC infrastructure, we conducted four case studies in four different domains: office (*Easymeeting*) (Resatsch et al., 2007), retail (*Mobile Prosumer*) (Resatsch et al., 2008), *Transport*

& *Ticketing* (Resatsch, 2010), and market research (*MUSE*) (Koenc et al., 2010):

- *Easymeeting* is an automatic "check-in" for office environments. With NFC tags attached to a wall, users can virtually check in at various meeting rooms and let their colleagues know where they are and how long the meeting room was blocked. This is synchronized with a calendar application.
- The *Mobile Prosumer* is a product information system using NFC tags attached to products. Touching a tag resolves the coded product ID, and based on the underlying Internet database, this ID is used to determine information about the product from professional sources and user opinions on the Internet.
- The *Transport & Ticketing* case uses a widespread NFC infrastructure in a bigger German city to let travelers purchase tickets for public transport with their NFC phones.
- The *MUSE* (Media Usage in Supportive Environments) application is a radio-frequency identification (RFID)-based prototype system that supports automatic measurements of print media usage in public environments. The MUSE prototype is designed for the logging of patients' print media consumption in a general practice anteroom. The utilization of the application follows the goal of delivering cost-effective and bias-free insights on user acceptance and consumption data to publishers. These data help publishers to optimize print media consumption among readers and increase value of advertisement space based on reader data.

The length of the research process in each case study—including the initial idea, prototype development, research design, evaluation, and final analysis—varied from 5 months (*Easymeeting*) to 10 months (*Mobile Prosumer, MUSE*), and to more than 18 months for the *Transport & Ticketing* case.

The *Easymeeting* (EM) application was developed in approximately two months. The case study entailed setting up the infrastructure, finding and convincing participants to take part in the study, obtaining the approval of management in two partner organizations, and designing the study. The actual execution of the study required one day in both organizations, including travel time. Another two months were required to analyze the questionnaires and the talking-out-loud method results.

The *Mobile Prosumer* (MP) case study took slightly longer, especially the phase that involved the organization of the case study. The setup of the case study—the research proposal, the study design and recruiting participants—required extra time. The actual focus groups were conducted within four hours. Transcription and analysis were completed in approximately two-and-a-half months.

The basis of the *Transport & Ticketing* (PTC) application was developed in advance of the actual case study. At least six months before the start of the evaluation, the working prototype was built, including the infrastructural components. The execution phase lasted four months with one screening point and three measurement points. The analysis required almost eight months to complete, mostly because of the sample size and the analysis program that was used.

The *MUSE* prototype application was designed for, and evaluated in, a Germany-based mid-sized medical practice anteroom as a typical environment for print media consumption. We wanted an application environment in which readers have the choice of a variety of journals and that is frequented by a demographically diverse reader group. Magazines were equipped with RFID tags and exhibited to individuals waiting for their treatment on a table equipped with an RFID transponder reading device. Data on the name and edition of a magazine were saved on RFID tags, which were attached to the magazines in order to facilitate a statistical analysis on media consumption, including name, type, reading time, and space. The prototype enables the fully automated logging of detailed reading duration data, which can be transformed to tabular and graphical outputs (e.g., pie and bar chart diagrams) through an integrated statistical component. The design development of the MUSE prototype took about six months and included qualitative interviews with print media publishers and medical practice staff in a requirements elicitation process. The evaluation of the MUSE prototype was conducted in a medical practice in a rural area near in Germany over the course of six weeks. During that time, approximately 1,200 patients visited the medical practice and spent some time, typically reading magazines in the waiting room. Analysis of the data collected by the

MUSE prototype was completed in approximately two-and-a-half months.

The idea behind the case studies was to find out more about people's *awareness* and *attitude* toward the technological potential of *NFC mobile phone and tag infrastructure*. The relevant findings are provided in this chapter in order to derive design guidelines for future NFC-based ubiquitous computing applications.

Nonfunctional Requirements for NFC Applications

The following set of requirements for ubiquitous computing environments enabled by an NFC infrastructure was derived in the analysis of the four foregoing case studies. They are meant for the creation of NFC applications on a nonfunctional requirement basis (Table 4.1).

Some of the less critical challenges might vary in different domains or industries. These challenges may also contradict one another and should be researched. For example, requiring full-coverage tagging of all objects may increase infrastructural start-up costs, thus increasing financial risks.

NFC Application Design Guidelines

Tables 4.2 to 4.6 list combined and overall findings from the four case studies in four areas: *NFC technology and application, tag infrastructure, human factors,* and *device*.

A better understanding of ubiquitous computing prototypes was achieved in the four case studies. Some of the findings applied to all case studies, while others were specific to one or two studies. Table 4.6 lists the most relevant findings and the case study in which they occurred in the four categories.

The established guidelines are a first starting point for designing NFC applications according to user needs and perceptions. They may help developers to better plan and design NFC applications. The recommendations for design guidelines for NFC applications may provide developers with an overview of important questions around NFC implementations, device specifics, tag placement, and human factors.

Table 4.1 Key nonfunctional requirements for NFC-based ubiquitous computing applications

REQUIREMENT	DESIGN GUIDELINE
Few choices after the single top-level choice.	NFC is a rather new concept. The technology easily confuses people if complex navigation structures are employed. The first nonfunctional requirement is based on Norman's (1999) notion of a shallow structure. In addition to the single top-level choice, only a few other function choices are important in terms of NFC tag usage. Every tag should trigger only one function or at least limit the functions shown on the mobile phone to a few options. Once NFC becomes well known, it might be possible to alter this.
Add general captions on labels.	Despite advances in logos and trademarks for NFC, the tag should clearly state which function it triggers in text form, symbols, or icons. These general captions should be placed on the tag and avoid the use of technical jargon. The RFID-specific word *tag* is itself largely unknown to people.
The user needs to immediately tell the state of the device and alternatives for actions.	With haptic feedback from the telephone, the user can clearly tell the current status of the information system. Haptic feedback should be turned on as standard. On the mobile front-end, any data transaction should be stated clearly with a progress bar or similar status indicators.
Relationship between actions and results must be clear and easy to determine.	NFC exists in the physical world, not only on a screen. This media break is difficult to understand for users. The relationship between tag (as initiator) and action (mobile service) should be clear insofar as the action suggested by the tag leads to the expected service. If that is not the case, confidence in the system will quickly be lost.
Tagging of context to information.	An NFC tag is related to a certain context. This can be a public space, an underground station, a park, a shop, etc. A tag needs to be situated and placed in a certain environment in which people are accustomed to using services. A tag for payments should be located either close to the cashier or in a process flow, for example, when leaving the sales floor.
Choice and consent: Provide a selection mechanism so that users can indicate which services they prefer.	The active behavior of touching a tag and starting a service implies certain consent. If there are more tags available at a site, the user can easily choose his or her service as a simple selection mechanism. Thus, offering the relationship of one service to one tag will not be realistically possible. Important, however, is establishing a close relation between choice and consent.

Table 4.2 NFC technology and application

FINDING	INTERPRETATION OF FINDING	DESIGN GUIDELINE
NFC is related to process efficiency.	NFC can be used to improve existing processes that require a greater degree of efficiency. These might include public transport ticketing, entry management, fleet management, parking tickets, etc., in which companies look to speed up process times in order to generate a larger number of transactions or to satisfy customers by reducing waiting times. The important relationship between NFC and processes implies that developers tailor the application to an existing process, not tailor the process according to the possibilities of NFC.	*Design for the process, not for NFC.*
An NFC application is usually considered very easy to use, but the wow-effect decreases quickly.	This finding has implications for marketing strategies. Campaigns targeting future consumers need to fulfill several NFC-related requirements: • This means that NFC ease of use should not be the sole focus of any marketing and promotional campaign. Ease of use drops in significance over periods of extended and sustained usage. • NFC tag processes or functions are extremely relevant and should not be overstated. Marketing should communicate that NFC makes a process faster or more convenient to the user (task performance). The wow-effect describes how stimuli tend to fade after repeated action. A communication of only the wow-effect would lead to misperceptions and frustration after longer usage.	*Focus in communicating the benefits of the application, not the technology.*

(Continued)

Table 4.2 NFC technology and application (Continued)

FINDING	INTERPRETATION OF FINDING	DESIGN GUIDELINE
Issuing party needs to be trustworthy.	In the transport and ticketing case study, participants expressed issues concerning NFC privacy and security. In the other studies, these issues were not relevant. Moreover, the average user is insecure about information provided via a mobile service. In particular, an NFC-based application provider should go to the necessary lengths to create a trustworthy image or to acquire the relevant certificates from issuing parties, etc.; in the NFC world, this is also conducted with Trusted Service Managers, at least for the payment area.	*Design trustworthy applications with clear terms of services for privacy and security.*
Implementation adapted to existing systems is a crucial factor.	The NFC application should be adapted to existing systems. Each company has fixed information system and application processes and existing information systems. NFC could be used to add on an easier entry point to the process. This requires valid frameworks and middleware to easily implement needed application interfaces, especially those concerning the tag infrastructure, the identification scheme, the back-end solution to transfer IDs to a service, etc. A customizable lightweight development framework is required.	*Design NFC to map existing systems.*
An NFC information system should be available for all products in a product group or all objects in an object group.	If tags are placed on multiple products and objects, it is crucial to provide the relevant services matching each product and object. Starting with a full implementation is recommended rather than providing only a few services. Each product, object, or location, and each tag should provide a specific service that the user wants or expects in a certain usage situation.	*Design relevant and specific services for all products, objects, or locations.*

Table 4.3 Tag infrastructure

FINDING	INTERPRETATION OF FINDING	DESIGN GUIDELINE
Tag placement is important in relation to the process.	Tag placement is one of the most important issues in NFC-based ubiquitous computing applications. NFC tags should be integrated as seamlessly as possible into current processes. NFC tags are small in scale, and thus hard to find if they do not have the right color, caption, or shape (see also tag shape). The evaluation should also test the placement in real-life situations. To increase and simplify usage and to increase adoption rates, maintain the same process and integrate NFC into the *flow*.	*Place the tags in the current process flow.*
Tag shape, caption, and placement are important in relation to the desired function of the tag.	The shape, caption, and placement should be easily identifiable in terms of what function is triggered by touching the tag. Because NFC is very new to the market, it is not clear to the average user what to do with an NFC tag. The touch-based interaction with a mobile phone is new. To avoid confusion, the industry needs to establish European-wide tag-appearance standards. Currently, only a few logos made available by organizations, such as the NFC Forum, exist; none of them have been applied widely. Depending on tag context, bright colors and self-explanatory symbols need to be added.	*Design a tag that is self-explanatory for the normal user (use captions, icons, colors).*
NFC tags should be available for all products in a product group or all objects in an object group.	Similar to the availability of information services, a tag should be easy to locate. This is important since NFC tag infrastructures are built around existing physical objects and, at times, within a limited amount of space. To facilitate usage, a tag has to be placed on all products, objects of a product, or object category to avoid disappointing customers if usage is only applicable in certain areas. Limited products, limited locations, and limited objects within a group featuring NFC tags will confuse users. Consequently, the user cannot adopt the service due to a lack of reliability when needed. The more tags there are and the easier they are to find, the better.	*Furnish all objects, locations, or products with tags.*

Table 4.4 Human factors

FINDING	INTERPRETATION OF FINDING	DESIGN GUIDELINE
The tagged target group is digital natives.	The digital natives target group responded best in two cases to several tagging scenarios. This group supposedly has an affinity for technology and is most likely to quickly adopt NFC-based services. More research needs to be done considering that digital immigrants also responded very positively to NFC in other studies.	*Design first for digital natives.*
Low impact of NFC on privacy and data security concerns.	Although privacy and security concerns were low, they need to be considered. However, participants' overall positive opinion of NFC indicates that companies should not hesitate to introduce NFC-based ubiquitous computing applications.	*See above.*
Simplicity is advantageous.	Simplicity is a very soft factor, but directly corresponds to application success. If NFC makes processes more efficient and greatly improves ease of use, simplicity will continue to convince users of the service. The conceptual model must be simple.	*Design simple applications.*
Low cognitive load improves understanding of NFC-based interaction.	Understanding the ubiquitous computing concept is not easy. To make ubiquitous computing applications useful, the application design needs to take a limited user understanding into account: The clear communication of what happens after a tag is touched is crucial to the service's acceptance. In keeping the cognitive load low, the guidelines, as shown by Norman (1999), help build the services.	*Design the application according to Norman's guidelines.*
NFC technology is best used for supporting existing processes.	Changing the process from its roots up is not the right way. NFC only aids in simplifying service initiation, rather than in initiating process changes. For example, in ticketing processes, NFC should not be introduced to change the ticketing process.	*Design for existing processes.*
Social influence is an important factor.	Social influence has an impact on user perceptions about new technologies. To create initial positive influences in adoption rates, it is important to either use a lead user strategy or viral marketing.	*Design with lead users.*

Table 4.5 Device

FINDING	INTERPRETATION OF FINDING	DESIGN GUIDELINE
Haptic feedback in a mobile device is considered positive for usage.	Haptic feedback after touching an NFC tag helped people understand when the transmission took place. An NFC-enabled phone should have haptic feedback turned on by default to assure users that touching the tag results in an action.	*Use haptic feedback as default.*
Existing device infrastructure needs to be considered.	Existing infrastructures are difficult to change. Market demand factors are needed to build NFC into phones.	*Build NFC in smartphones first.*
NFC reader location needs to be clearly identifiable.	Study participants had problems locating the antenna when using the phone for the first time. Even if an introductory explanation helped people in this regard, users who do not have the benefit of expert instruction must be able to easily identify the reader's location. With simple touch symbols, first usage becomes easier and the barrier to trying services decreases.	*Clearly indicate where the reader is placed.*

Table 4.6 Findings and number of occurrences in conducted case studies

FINDING	EM	MP	PTC	MUSE	DESIGN GUIDELINE
(A) NFC TECHNOLOGY AND APPLICATION					
NFC is related to process efficiency.	X		X	X	*Design for the process, not for NFC.*
An NFC application is usually considered very easy to use, but the wow-effect decreases quickly.	X	X	X		*Focus on communicating the benefits of the application, not the technology.*
Issuing party needs to be trustworthy.		X		X	*Design trustworthy applications with clear terms of services for privacy and security.*
Implementation to existing systems is a crucial factor.	X	X	X		*Design NFC to map existing systems.*
An NFC information system should be available for all products in a product group or all objects in an object group.		X	X		*Design relevant and specific services for all products, objects, or locations.*
(B) TAG INFRASTRUCTURE					
Tag placement is important in relation to the process.	X	X	X	X	*Place the tags in the current process flow.*
Tag shape, caption, and placement are important in relation to the desired function of the tag.	X		X		*Design a tag that is self-explanatory for the normal user (use captions, icons, and colors).*
NFC tags should be available for all products in a product group or all objects in an object group.		X	X	X	*Furnish all objects, locations, or products with tags.*
(C) HUMAN FACTORS (NONFUNCTIONAL REQUIREMENTS)					
The tagged target group is digital natives.		X			*Design first for digital natives.*
Low impact of NFC on privacy and data security concerns.	X	X			*See above.*

Table 4.6   Findings and number of occurrences in conducted case studies

FINDING	EM	MP	PTC	MUSE	DESIGN GUIDELINE
Simplicity is advantageous.	X				*Design simple applications.*
Low cognitive load improves understanding of NFC-based interaction.	X	X			*Design the application according to Norman's guidelines.*
NFC technology is best used for supporting existing processes.	X		X	X	*Design for existing processes.*
Social influence is an important factor.	X		X		*Design with lead users.*
(D) DEVICE					
Haptic feedback in a mobile device is considered positive for usage.	X	X			*Use haptic feedback as default.*
Existing device infrastructure needs to be considered.	X	X	X		*Build NFC in smartphones first.*
NFC reader location needs to be clearly identifiable.	X	X			*Clearly indicate where the reader is placed.*

The challenges for developers/companies discussed in this study are (1) the justification of high initial investments in infrastructure, (2) unknown user technology acceptance, (3) unknown technology among future users (RFID/NFC), and (4) the need of consumers and users to experiment with NFC to determine its values. To overcome these barriers and challenges, the established guidelines should support developers and companies in creating applications and evaluating them accordingly in the early stages.

Guideline Evaluation with the NFriendConnector

The NFriendConnector prototype enables seamless integration of users' off-line social interactions with their online social networks using a mobile phone. It allows users to instantaneously initiate and establish Facebook connections using their mobile phones without having to incur any additional search cost and therefore supports the usage behavior of "social searchers" (Lampe et al., 2006). The general underlying use case is rather simple:

> Two individuals that are members with an existing user profile on the social computing platform Facebook meet in a real-life/off-line situation in which human–human interaction occurs, such as in a club, café, park, or comparable socializing situation. Both individuals are carrying a cellular device with an available mobile Internet connection. After establishing a physical contact with their mobile phones, they can now enrich their socializing situation by viewing and comparing the Facebook profile data exchanged through the NFC interface connection between the two mobile devices, and map it onto the Facebook platform by creating a new Facebook friend-connection or status message.

Due to the usage of NFC technology in the NFriendConnector application, an interaction and exchange of information between users has to be initiated by the users themselves by establishing a near-physical contact between the two mobile devices. The application is not doing things in the background without the users' active participation or knowledge. Since the user has full control over the exchange of information that happens with other users, the NFriendConnector overcomes the privacy concerns associated with other mobile social networking applications. The NFriendConnector prototype intends to fill the significant gap between the initiation of contact in a real-life social setting and the online social network. Given the ubiquitousness of mobile phone usage in today's world, we believe that mobile phones have the significant potential of lowering the burden involved in establishing social networking connections, especially for people who frequently use Facebook to establish online connections with their off-line contacts. Once the social networking profile information is present on mobile phones, it can further enrich the actual face-to-face interaction experience by enabling users to view, share, and match information regarding themselves and others.

The prototype provides a simple set of functionalities, partitioned into client-enabled and connection-enabled, to support the foregoing off-line socializing situation use case. Client-enabled functionalities (1) *view profile*, (2) *match profile*, and (3) *save profile* of the NFriendConnector prototype application can be executed by the user on her mobile device without an active mobile Internet connection. In that case, the application accesses data stored on the mobile device itself and is updated whenever the user logs into the Facebook platform

Figure 4.1   (Left) "Mode Options" menu and (right) "Profile Options" menu of NFriendConnector.

through the NFriendConnector prototype application. Additional, connection-enabled functionalities (4) *add as friend* and (5) *make status message* can only be executed by the user while logged into the Facebook platform through the NFriendConnector prototype application on a constant mobile Internet connection. The user can choose between these two modes upon the start sequence of the application, and can switch modes later, should the usage situation change. The application starts with a splash screen that leads the user to the *mode options* menu, as displayed in Figure 4.1.

After choosing either the connected or nonconnected mode of NFriendConnector, the user can initiate an NFC communication by touching another NFC-enabled mobile phone running NFriendConnector, which will present him with the *Profile options* menu (as seen in Figure 4.1) to access the functionalities described in the following section. An overview of the requirements, modes, and features of the NFriendConnector prototype application is given in Table 4.7.

In the next section, we will demonstrate the adoption of the introduced requirements and design guidelines for NFC application development and their influence on design decision during the conceptual design and implementation of the NFriendConnector prototype. Each requirement or design guideline is transformed into a design rationale for the conceptual design of the NFriendConnector, where applicable. If a requirement

Table 4.7 Requirements, modes, and features of NFriendConnector client-enabled functions

NFRIENDCONNECTOR		
Requirements	NFC-enabled mobile device	
	Existing Facebook account with profile information	
		Available mobile Internet connection
Modes	Stand-alone	Connected
Features	View Profile	
	Match Profile	
	Save Profile	
		Add as friend
		Make status message

or design guideline is not applicable to the NFriendConnector prototype, we simply omit it in the following listing.

Additionally, at the end of this section, we present findings on user reactions to the NFriendConnector prototype, elaborated in a laboratory experiment that allows users to test the application. It was particularly interesting to note that even people who had no prior experience with NFC or Facebook perceived the prototype as being equally useful and were equally satisfied with it (reflected by no significant difference in the mean values of perceived usefulness or satisfaction for the Facebook user and Facebook nonuser group).

The following nonfunctional requirements were transformed into corresponding design rationales and applied during the NFriendConnector prototype development process:

- *Few choices after the single top-level choice:* The NFriendConnector prototype provides the user with an icon-based navigation (see Figure 4.1) following the notion of a shallow structure (Norman, 1999). Users can access all major functionalities through a single top-level choice in the navigation. Only the (4) *add as friend* functionality is realized in a more complex fashion due to the Facebook application programming interface (API), which can only be accessed through a mobile browser.

- *The user needs to immediately tell the state of the device and alternatives for actions:* The supported course of action within the described use case scenario envisions a single NFC connection between two mobile phones. We therefore implemented a short message that informed the user of a positively established NFC connection and termination of profile data transmission.

- *Relationship between actions and results must be clear and easy to determine:* The NFriendConnector prototype *use case* scenario builds on a single NFC connection and data transmission. After the "handshake," the prototype informs the user with a short message whether the data transmission is positive or interrupted. After the transmission, the provided functionalities are uncoupled from the NFC technology itself.

In the next step, we outline the adoption and influence of established NFC application design guidelines (see Tables 4.2–4.5) during the NFriendConnector prototype development:

- *Design for the process, not for NFC:* The NFriendConnector offers Facebook users the possibility of creating Facebook connections, and accessing other Facebook functionalities at the point of their real-life social interaction, using their mobile devices. Consequently, the prototype supports the rather complex process of connecting to social ties in online social networks in real-world situations.

  The current process is rather complex and can be described as follows:

  In order to establish a connection with a person one has met in some kind of off-line social setting such as a common class, a party, or a conference, one has to go back to the Facebook account and search for the other person's profile. Once found, a request to become friends has to be sent to the other person and afterward accepted by him. Establishing connections in this form involves certain costs to the person initiating the connection in the form of having to search for the other person's profile, waiting for the other person to respond and, above all, having access to a computer terminal with Internet connectivity. The NFriendConnector prototype supports this off-line process by mimicking a physical "handshake" that is assisted by the application of NFC technology. All functionality of the prototype only depends on a single NFC connection and data transmission.

- *Design trustworthy applications with clear terms of services for privacy and security:* In contrast to mobile social networking application prototypes, which are mainly based on Bluetooth sensing, the NFriendConnector does not use sensory data such as physical proximity or location to automatically initiate a profile matching or the generation of usage data. Due to the restriction to NFC technology for data transmission in the NFriendConnector prototype, an interaction and exchange of information between users has to be initiated by the users themselves by establishing a near-physical contact between the two mobile devices (e.g., physical "touch" between the devices). The system is not processing or exchanging data in the background without the users' active participation or knowledge. As previously noted, since the user has full control over the exchange of information with other users, the NFriendConnector overcomes the privacy concerns associated with other mobile social networking applications. In this case, NFC enables a trustworthy environment for the user.

The previously introduced guidelines on tag infrastructure are not applicable for the NFriendConnector prototype since the application is not based on the utilization of NFC tags. Further, we demonstrate the application and influence of previously introduced guidelines which consider human factors:

- *Design simple applications:* The NFriendConnector application is limited to a single-layer navigational structure and five key functionalities in order to keep it simple and easy to operate on mobile devices with limited display measures and input options.
- *Design the application according to Norman's guidelines:* We decided to build the NFriendConnector prototype with an icon-based navigation (see Figure 4.1) following the notion of a shallow structure (Norman, 1999). A potential user can access all major functionalities through a single top-level choice in the navigation; only the "add friend" functionality is realized in a more complex fashion due to the Facebook API, which can only be accessed through a mobile browser. All icons are supported with a textual description of the

represented functionality. The choice of icons is based on commonly used icons for general functionalities, for example, save (profile) data.

We could only apply one of the three design guidelines concerning devices equipped with NFC technology, since the NFriendConnector application and its supported processes are not based on an NFC tag infrastructure. We therefore want to outline our design rationale for the following:

- *Haptic feedback in a mobile device is considered positive for usage:* The supported course of action within the described use case scenario envisions a single NFC connection between two mobile phones. We therefore refrained from haptic feedback and instead implemented a short message that informed the user of a positively established NFC connection and termination of profile data transmission.

  Adding additional haptic feedback after touching the mobile phones could help people to understand when the transmission of profile data took place. We decided against additional haptic feedback on the premise that the vibration could degrade the "handshake" metaphor.

The primary goal of any prototype development exercise and evaluation is that of addressing a particular user need that is not currently addressed and therefore improving users' overall experience of using an application or system. An experimental methodology was chosen to assess the usability of the NFriendConnector prototype. Subjects were recruited from volunteers in a large university in Germany. The subjects were mostly registered in the university's undergraduate or graduate programs. The experiment was held over multiple sessions in one particular room that was designated as the experimental laboratory. Each experimental session lasted for about 30 minutes. A standard protocol was followed for all the sessions. Subjects were asked to complete a short questionnaire about their online social networking behavior and experience, before being provided with a brief introductory description about the NFriendConnector prototype explaining the supported use case and its various features. They were then given a demonstration of how they could use the prototype. Following the demonstration, they were provided with NFC-enabled mobile phones that had the prototype installed on it, and were asked to evaluate the prototype in a follow-up questionnaire. In order to assess the usability of the prototype, each subject was paired with another subject, and they were then asked to use the NFriendConnector prototype application for establishing Facebook connections with each other. It was particularly interesting to note that even people who had no prior experience with NFC and Facebook perceived the prototype as equally useful, and they were equally satisfied with it (reflected by no significant difference in the mean values of perceived usefulness or satisfaction for the user and nonuser group).

We conclude that the introduced nonfunctional requirements and design guidelines helped us develop a streamlined application that could be used intuitively by all subjects, despite their lack of previous experience with NFC technology. Through the application of a shallow structure, subjects were able to immediately tell the state of the device and alternatives for actions. We could also observe that subjects understood the relationship between actions and results of provided functionalities, as they continued to test the prototype after completing the test protocol.

We therefore assume that the previously introduced nonfunctional requirements and guidelines helped us in the conceptual design and development of a user-friendly and potentially successful NFC application.

Limitations and Future Research

Our conducted case studies, the derived and compiled nonfunctional requirements and design guidelines, as well as their application within the NFriendConnector prototype design process should be interpreted in the context of applicable limitations. Our derived nonfunctional requirements and design guidelines originated from a limited set of case studies. Future case studies could either generate additional nonfunctional requirements and guidelines applicable for the development of NFC applications, or reinforce the set of introduced requirements and guidelines. The design guidelines presented in this chapter are a first start to actually introducing the NFC technology and NFC-enabled applications to markets, and for understanding how NFC will gain momentum during the next few years. The design guidelines therefore need to be tested to check how they influence NFC adoption. Future research endeavors may be targeted toward enhancing the generalizability of presented nonfunctional requirements and design guidelines by applying them to future NFC-enabled prototype applications.

Remarks

Parts of this contribution have been previously published with Gabler (Resatsch, 2010); please also see the original source.

The NFriendConnector prototype was developed in the context of the research project Mobil50+ (Innovative NFC- und IT-basierte Dienstleistungen für mobiles Leben und Aktivität der Generation 50+). Mobil50+ is funded by the German Federal Ministry of Education and Research (BMBF - FKZ: 01FC08046). It is a joint project of the Technische Universität München and various partners. For further information, see www.projekt-mobil50.de.

References

Bleek, W.-G., Jeenicke, M., and Klischewski, R. (2002). *Developing* web-based applications through e-prototyping. *Proceedings of the 26th International Computer Software and Applications Conference on Prolonging Software Life: Development and Redevelopment*, Oxford.

Blomberg, J. L. and Henderson, A. (1990). Reflections on participatory design: Lessons from the trillium experience. *SIGCHI Conference on Human Factors in Computing Systems: Empowering People*, Seattle, Washington.

Bødker, S., Grønbæk, K., and Kyng, M. (1995). Cooperative design: techniques and experiences from the Scandinavian scene *Human-computer Interaction: Toward the Year 2000* (pp. 215–224). San Francisco, CA: Morgan Kaufmann.

Buchenau, M. and Suri, J. F. (2000). Experience prototyping. *DIS 2000, Third Conference on Designing Interactive Systems: Processes Practices, Methods, and Techniques*, Brooklyn New York.

Davis, F. D. and Venkatesh, V. (2004). Toward preprototype user acceptance testing of new information systems: implications for software project management. *IEEE Transactions on Engineering Management, 51*(1), 31–46.

Floyd, C. (1984). A systematic look at prototyping. *Approaches to Prototyping*, 1–18.

Floyd, C. (1989). Softwareentwicklung als Realitätskonstruktion. *Informatik-Fachberichte, 212*, 1–20.

Gourville, J. T. (2006). Wann Kunden neue Produkte kaufen. *Harvard Business Manager*, August, 45–53.

Greenbaum, J. and Kyng, M. (1992). *Design at work: cooperative design of computer systems*: Lawrence Erlbaum Associates. Mahwah, NJ.

Gruner, K. E. and Homburg, C. (2000). Does customer interaction enhance new product success? *Journal of Business Research, 49*(1), 1–14.

IEEE (1990). IEEE Standard Glossary of Software Engineering Terminology Retrieved 19.07.2010, from http://standards.ieee.org/reading/ieee/std_public/description/se/610.12-1990_desc.html

Koene, P., Köbler, F., Burgner, P., Resatsch, F., Sandner, U., Leimeister, J. M. et al. (2010). *RFID-based Media Usage Panels in Supportive Environments*. ECIS 2010 - 18th European Conference on Information Systems, Praetoria, South Africa.

Kotonya, G., Sommerville, I. (1998). *Requirements Engineering: Processes and Techniques*. Chichester, UK: Wiley.

Lampe, C., Ellison, N., and Steinfield, C. (2006). A face(book) in the crowd: Social searching vs. social browsing. *CSCW 2006 20th Anniversary Conference on Computer Supported Cooperative Work*, Banff, Alberta.

Morrison, P. D., Roberts, J. H., and Midgley, D. F. (2000). Opinion leadership amongst leading edge users. *Australasian Marketing Journal, 8*(1), 5–14.

Muller, M. J. (1993). PICTIVE: Democratizing the dynamics of the design session. *Participatory Design: Principles and Practices*, 211–237.

Muller, M. J. (2003). *Participatory Design: The Third Space in HCI*. Mahway, NJ: Erlbaum.

Muller, M. J., Blomberg, J. L., Carter, K. A., Dykstra, E. A., Madsen, K. H., and Greenbaum, J. (1991). *Participatory design in Britain and North America: Responses to the "Scandinavian Challenge"*. SIGCHI conference on Human factors in computing systems: Reaching through technology, New Orleans, Louisiana.

Nambisan, S. (2002). Designing virtual customer environments for new product development: Toward a theory. *The Academy of Management Review, 27*(3), 392–413.

NFCForum (2007). Near Field Communication in the real world: Turning the NFC promise into profitable, everyday applications Retrieved 16.07, 2010, from http://www.nfc-forum.org/resources/white_papers/Innovision_whitePaper1.pdf.

Norman, D. A. (1999). Affordance, conventions, and design. *Interactions, 6*(3), 38–43.

Resatsch, F. (2010). *Ubiquitous Computing: Developing and Evaluating Near Field Communication Applications*: Gabler, Betriebswirt.-Vlg.

Resatsch, F., Karpischek, S., Sandner, U., and Hamacher, S. (2007). Mobile sales assistant: NFC for retailers. *Mobile HCI 2007 9th International Conference on Human–Computer Interaction with Mobile Devices and Services*, Singapore.

Resatsch, F., Sandner, U., Leimeister, J. M., and Krcmar, H. (2008). Do point of sale RFID-based information services make a difference? Analyzing consumer perceptions for designing smart product information services in retail business. *Electronic Markets, 18*(3), 216–231.

Rukzio, E., Leichtenstern, K., Callaghan, V., Holleis, P., Schmidt, A., and Chin, J. (2006). An experimental comparison of physical mobile interaction techniques: Touching, pointing and scanning. *UbiComp 2006 Eighth International Conference on Ubiquitous Computing Orange County*, California.

Simon, H. A. (1980). The newest science of the artificial. *Cognitive Science*, 4(1), 33–46.

Simon, H. A. (1996). *The Sciences of the Artificial* (3 ed.). Cambridge, MA: MIT Press.

Urban, G. L. and von Hippel, E. (1988). Lead user analyses for the development of new industrial products. *Management Science, 34*(5), 569–582.

von Hippel, E. (1986). Lead users: A source of novel product concepts. *Management Science, 32*(7), 791–805.

von Hippel, E. (1988). *The Sources of Innovation*: New York: Oxford University Press.

von Hippel, E. (2001). User toolkits for innovation. *Journal of Product Innovation Management, 18*(4), 247–257.

von Hippel, E. (2005). *Democratizing Innovation*. Cambridge, MA: MIT Press.

5

SOFTWARE SUPPORT FOR THE USER-CENTERED PROTOTYPING OF MOBILE APPLICATIONS THAT APPLY NEAR FIELD COMMUNICATIONS

KARIN LEICHTENSTERN AND ELISABETH ANDRÉ

Contents

Introduction

When developing a mobile application, the utilization of software support can improve the resulting product as well as the development process by improving cost-effectiveness and time-effectiveness (Myers 1995). In this chapter, we cover different approaches to software support for developers of mobile applications that apply Near Field Communications (NFC) in the context of the *Third Paradigm*, and compare two of these software approaches in a user study. In particular, we address software that assists in the user-centered development process of mobile applications (Kangas and Kinnunen 2005; Mao et al. 2005). A characteristic feature of this process is iterative prototyping, which includes several iterations of designing interface prototypes along with continuous evaluations and analyses of these prototypes with end users. Consequently, in this chapter, we not only cover software support that assists the design and implementation of mobile applications but also in their evaluation and analysis.

The Third Paradigm

NFC in the context of mobile payment is well known and a hot topic in industry. For instance, the Deutsche Bahn makes use of mobile phones with built-in NFC readers in their project, called *Touch&Travel*. In this project, users of the Deutsche Bahn can apply their mobile phones to pay their train tickets. They simply need to touch an NFC access point with their phone to indicate their departure and arrival station. Afterward, the ticket price is automatically charged. NFC, however, can not only be applied for mobile payment but more generally in the context of the so-called *Third Paradigm*: *The Internet of Things*, *Pervasive* or *Ubiquitous Computing*.

The idea of the Third Paradigm is to enable interactions with physical objects of the user's everyday environment, such as an object of art in a museum (Weiser 1991; Kindberg et al. 2000; Mattern and Floerkemeier 2010). Users can immediately interact with physical objects in the real world that incorporate computerized technology (e.g., an interactive surface) or make use of an interaction device as a medium in order to indirectly interact with them. In the context of the Third Paradigm, the primary interaction devices are mobile phones. Almost everybody owns a mobile phone and carries it around constantly. Recent phones support novel hardware and network facilities that enable different mobile interactions with physical objects. Ballagas et al. (2006) give a comprehensive overview of the hardware facilities that enable mobile interaction techniques with modern smart phones. For instance, the built-in camera or NFC reader of a mobile phone can be used to enable interactions with physical objects to receive services such as a detailed description of the object (Kindberg et al. 2000). For the NFC interaction, the physical object needs to be augmented with an RFID tag that provides information about the object's identifier. Now, once the phone is within a short distance to an RFID tag, an NFC connection is built, and the identifier is transmitted to the mobile phone. This identifier can then be used for various applications, such as database requests to receive additional information about the physical object (e.g., the artist of an artwork).

User-Centered Prototyping

Several mobile applications in the context of the Third Paradigm have been developed in order to investigate end-user preferences and behavior (Rukzio, Leichtenstern et al. 2006; Rukzio, Broll et al. 2007; Broll et al. 2009). For instance, in a former work (Rukzio, Leichtenstern et al. 2006), we investigated the user preference for a mobile interaction technique when using a mobile phone as a remote control for physical objects (e.g., DVD player, heating) in a smart flat. In this work, we first implemented a mobile application that applied different hardware facilities, including the NFC reader, to support three mobile interaction techniques. Later, we conducted a user study with end users. In this study, the subjects ran through different contextual situations (e.g., different distances to the intended object). Finally, we analyzed the user study and established guidelines regarding which technique is favored in which situation. In another former work (Leichtenstern and André 2009), we used mobile phones as interaction devices in a pervasive multiuser game called the *World Explorer*. In this game, three children could jointly select physical objects that represented different countries and corresponding topics. After the selection, the children could answer questions about the intended country and topic, such as questions about the geography and history of Australia. In this work, we were interested in the user behavior when assigning the interaction devices in three different modalities. We investigated whether the level of activity of each group member and social interactions can be increased as well as whether off-task behavior can be reduced in one of the three modalities.

In the first modality, each child was equipped with a single mobile phone, whereas in the second modality only one phone was available per group. In the last modality, each group member owned a separate mobile phone, but each of these devices only supported a subset of the entire interaction spectrum that was available in the other two modalities. Thus, only by using all three mobile phones together could the game be controlled and the task successfully completed. In this work, again, we had to put in a lot of effort to first design and implement different prototypes for the mobile phones, then to run a user study, and finally to analyze several hours of the captured videos in order to reveal the user behavior in the different modalities. But not just in research, in industry too we more and more see the development of applications involving end users in user studies. By this means, applications can be developed to fulfill several user requirements and the resulting products can support a good design and usability, which is an important quality criterion and competitive advantage.

In sum, the development of mobile applications with the involvement of the end user is more and more a commonly used approach that demands the following phases: the design and implementation as well as the evaluation and analysis of prototypes. This pattern corresponds to the iterative prototyping of the ISO norm 13407's description of the user-centered design process. The description claims a process that will produce design solutions until the product meets all user requirements for being as user-friendly as possible (Kangas and Kinnunen 2005; Mao et al. 2005). The development of mobile applications in those user-centered prototyping iterations, however, can often be very time consuming and expensive, especially if the developers are less skilled in software and usability engineering. For instance, the developers of pervasive interfaces for mobile phones require knowledge of network programming and of how to address different built in hardware, such as the microphone, camera, GPS interface, and NFC reader. Additionally, interface developers need to know how to deal with limited input and output functionalities (e.g., small displays). Finally, they also require knowledge of approved interface guidelines as well as experience in the conduct and analysis of user studies in order to get meaningful results. As a consequence, Myers (Myers 1995) claims the need for software support to overcome these challenges.

In the remaining parts of this chapter, we first describe different types of software support for mobile application development as well as software tools that support the different phases of user-centered prototyping. Then, we describe a tool that supports all stages of user-centered prototyping of mobile applications and, finally, its evaluation to validate benefits and problems when applying it compared to traditional approaches.

Software Support for Mobile Application Development

Based on Hull, Clayton, and Melamed (2004), developers can employ different types of software support when user-centered prototyping mobile applications: middleware or software modules as well as sketching or content and behavior tools. These four categories, however, primarily support the design and implementation of applications, but do not additionally focus on support for user evaluations and analyses. To support assistance for user-centered prototyping in all stages, approaches are required that also assist the conduct and analysis of user evaluations. Thus, we added two further types of tools to Hull's categories: evaluation and analysis tools as well as user-centered prototyping tools.

Middleware and Software Modules

Eclipse and Netbeans are two well-known Integrated Development Environments (IDEs) that can be used for the software implementation of mobile applications. These IDEs make use of so-called emulators of mobile phones (e.g., from Nokia or Sony Ericsson) in order to enable emulations and tests of mobile applications on a desktop computer during the development process. Typically, these emulators provide a set of different so-called Application Programming Interfaces (APIs). APIs typically provide software access to different functionalities of a mobile phone. For instance, most modern emulators of J2ME (Java Microedition) support the *JSR 135 Mobile Media API* that provides interfaces and classes to address the mobile phone's built-in camera and microphone as well as to run audio and video files. The *JSR 257 Contactless Communication API* is an additional API that is required whenever a mobile application is developed that is intended to support NFC. Most emulators do not provide this NFC API in their standard installation. Additional APIs such as the NFC API, however, can be downloaded (e.g., from the Nokia developer website) and added to a project of an IDE.

Middleware and *software modules* are also software components that typically provide an API to access different functionalities, such as access to different hardware capabilities of a device or a network layer. Usually, they also can be added to a project of an IDE. Equip (Greenhalgh 2002) is an example of a middleware that supports developers of mobile applications. It offers a software platform for mixed-reality user interfaces. In contrast to middleware, modules provide software for different components of software architectures, such as a server component. Using software modules, developers do not need to reimplement the corresponding component anymore but instead can reuse it in different contexts. For instance, the *Context Toolkit* (Salber, Dey and Abowd 1999) provides software to address different sensors.

Despite these benefits, middleware and software modules, however, seem to insufficiently meet the requirements of user-centered prototyping. They provide developers with valuable support when developing applications that can save time and money as well as prevent errors, but even if they are used in combination with

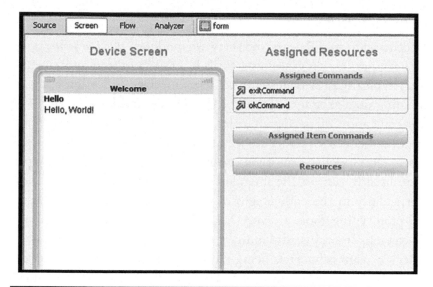

Figure 5.1   Screen view of the GUI builder from Netbeans for the development of mobile applications' appearance.

an IDE that provides a graphical user interface (see Figures 5.1 and 5.2), comprehensive programming skills are still required. Additionally, middleware and software modules typically just support the design and implementation and not the evaluation and analysis of applications.

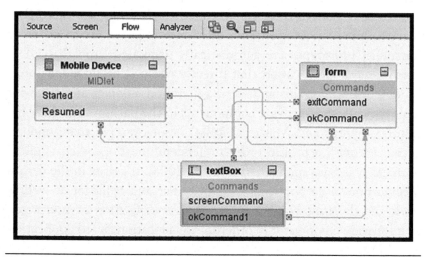

Figure 5.2   Flow view of the GUI builder from Netbeans for the development of mobile application's behavior.

Classical Prototyping Tools

In contrast to middleware and software modules, *classical prototyping tools* require less programming skills but fail in other requirements. They typically provide a graphical user interface and often assist in building up a sketch of a concept (Li, Hong, and Landay 2004) or a specification of particular rules, for example, for context-aware applications (Dey et al. 2004), but they provide assistance in generating low-fidelity prototypes rather than high-fidelity prototypes that directly run on the intended devices, such as mobile phones. Particularly in the early stages of the development process, classical prototyping tools are very helpful in communicating ideas for discussions in an interdisciplinary project team. We, however, consider software support as being more feasible as it also helps generate more functional prototypes that directly run on the intended interaction devices.

Content and Behavior Tools

All three introduced categories of tools fail to assist the evaluation and analysis of prototypes, which is also applicable to content and behavior tools. They only assist whenever an application has to be generated. A GUI, however, is provided for the generation of high-fidelity prototypes. Usually, the developers just need to specify the content and behavior of the intended application via this GUI. No or minimal implementation skills are required anymore, since functional prototypes are automatically generated from this GUI-based design specification that directly run on the intended interaction devices. For this automatic generation, content and behavior tools typically use software components in the background. MScape (Hull, Clayton, and Melamed 2004) is an example that supports developers with a tool in order to generate location-based applications, such as Savannah (Benford et al. 2004). It assists in the design specification of an application's appearance and behavior as well as static and dynamic content. This XML specification is interpreted by a software component in order to run a location-based application on a PDA.

Evaluation and Analysis Tools

Evaluation and analysis tools assist in the conduct and analysis of user evaluations. These tools also fail to meet all the requirements of user-centered prototyping because the design and implementation of prototypes is not supported. They typically provide a GUI to record user studies while subjects interact with an application. Later, the tools assist in analysis of these captured data. For instance, Momento (Carter, Mankoff, and Heer 2007) supports developers in the automatic logging of user interactions and the audiovisual capturing of user studies. After the conduct of the user study, Momento additionally assists in the analysis of user studies by providing a GUI that displays the captured video, which is automatically annotated based on the logged user interactions. Using this assistance, developers can quickly and easily find and jump to relevant sections in the videos, for example, to find usability problems or investigate the user's behavior.

User-Centered Prototyping Tools

In conclusion, content and behavior tools support the design and automatic implementation of high-fidelity prototypes while evaluation and analysis tools support the conduct and analysis of user studies. In combination, they can cover all stages of user-centered prototyping. These tools, however, are often not compatible, and therefore the smooth progress of the different phases can be interrupted. Moreover, developers need to learn different tools with different features and interfaces, which can waste development time. *User-centered prototyping (UCP) tools* support developers with an all-in-one-software in all stages of user-centered prototyping (Hartmann and Klemmer 2006; Klemmer et al. 2000). (1) An all-in-one UCP tool assists developers in the tool-based design specification, evaluation, and analysis of interface prototypes. Usually, UCP tools feature a strong link between the design, evaluation, and analysis components, which can prevent interruptions of the process; for example, the evaluation component supports the conduct of user studies with prototypes that were generated during the tool-based design and thus provides a logging mechanism for the user studies. The analysis component assists designers with the interpretation of synchronously captured data via the evaluation component. (2) A further feature of UCP tools is the generation of evolutionary prototypes with a high fidelity. In terms of interface design, a prototype represents a partial simulation of an interface with respect to its final appearance and behavior (Houde and Hill 1997). Evolutionary prototypes are characteristically prototypes that are built in rapid iterative development cycles in order to experimentally find and validate the user's requirements. In each cycle, evolutionary prototypes are modified and tested with experts and end users until they meet all requirements of the end users (Davis, Bersoff, and Comer 1988; A. Davis 1992). A UCP tool should generate evolutionary prototypes that provide several implemented functionalities of the final product and directly run on the respective interaction devices, for example, mobile phones. These high-fidelity prototypes rather enable realistic user studies and prevent user misunderstanding of the interface in user studies (Holmquist 2005) than mock-ups of an interface (low-fidelity prototypes) that simulate several functionalities. (3) A last important aspect of UCP tools is the support for remote usability studies. A remote usability study characteristically means a spatial separation of the subjects and evaluators (Andreasen et al. 2007) during a user study in a laboratory or in a field setting (in situ, that is, at home or at the office). Typically, in a remote usability study, the user interactions (e.g., mouse clicks) are logged while the user is audiovisually captured. Moreover, screen shots or records are saved from the interface (e.g., Dow et al. 2005).

User-Centered Prototyping Tools

Software tools seem to provide several benefits for user-centered prototyping. Most importantly, they provide a GUI in order to reduce the required software and usability engineering skills. Now, we illustrate tools for the design, evaluation, and analysis of prototypes, and then based on these tools we introduce and explain typical supported features for developers in user-centered prototyping.

Tool Support for User-Centered Prototyping

Most available tools support just one of the three described phases (design, evaluation, and analysis). Consequently, these tools are rather content and behavior tools or evaluation and analysis tools than user-centered prototyping tools. Nevertheless, all the following tools can provide valuable inputs on feasible tool features for a user-centered prototyping tool.

Tools for Design   MakeIT (Holleis and Schmidt 2008), OIDE (McGee-Lennon et al. 2009), MScape (Hull et al. 2004), and TERESA (Chesta et al. 2004) are examples of tools that just support the design of a high-fidelity prototype. MakeIT and MScape support developers of pervasive or ubiquitous interfaces. TERESA addresses the tool-based design of functional nomadic interfaces and OIDE the generation of multimodal interfaces. The mixed-fidelity prototyping tool from Sá (de Sá et al. 2008) supports the design and conduct of user studies for mobile phones.

Tools for Prototype Evaluation and Analysis   Compared to these tools, there are fewer tools that assist in the evaluation and analysis of user evaluations. Most available *evaluation and analysis tools* support the logging of web traffic and their visualization (e.g., Arroyo et al. 2006). MyExperience (Froehlich et al. 2007), DRUM (Macleod and Rengger 1993), Momento (Carter et al. 2007), and UMARA (Bateman et al. 2009) are examples of tools that support the conduct of user studies or their analyses. MyExperience supports the recording of evaluations for mobile phone applications, whereas DRUM, UMARA, and Momento are both evaluation and analysis tools that even concentrate on tests for mobile (Momento) or desktop settings (DRUM and UMARA).

Tools for User-Centered Prototyping   The only known tools that support all three process steps—the dynamic generation of interactive evolutionary high-fidelity prototypes and the conduct of local and remote usability studies—of user-centered prototyping in a single software are D.tools (Hartmann and Klemmer 2006), SUEDE (Klemmer et al. 2000), and MoPeDT (Leichtenstern and André, The Assisted User-Centred Generation and Evaluation of Pervasive Interfaces 2009; Leichtenstern and André, MoPeDT—Features and Evaluation of a User-Centred Prototyping Tool 2010). Klemmer and colleagues developed SUEDE, which assists in the iterative development of speech interfaces, whereas Hartmann and colleagues implemented D.tools, which supports the design, evaluation, and analysis of physical computing applications. SUEDE is used to design dialogue examples, evaluate the examples in a Wizard of Oz setting and, later on, to analyze the evaluation, such as the dialogue path chosen by the user during the tests. D.tools can be primarily applied to develop, test, and analyze new information appliances in a laboratory, such as new media players or cameras and their buttons and sliders. MoPeDT offers a wide-ranging playground to conduct different user studies in pervasive environments with mobile phones as interaction devices. Interface developers can design applications in the context of the Third Paradigm (e.g., a pervasive shopping assistant), locally or remotely evaluate the application in the pervasive environment and, later on, analyze the results. MoPeDT is the only known tool that supports developers during the user-centered development of interactive, evolutionary, high-fidelity prototypes for mobile phones in order to cope with the previously mentioned challenges, such as the comprehensively required programming and interface skills. Using MoPeDT, applications for mobile phones can be generated that support different pervasive interaction techniques (Leichtenstern and André, The Assisted User-Centred Generation and Evaluation of Pervasive Interfaces 2009) including a technique based on NFC for the interaction with physical objects (e.g., products in a store or objects of art in a museum).

Tool Feature for User-Centered Prototyping

After having introduced examples of software tools, we now describe typical supported tool features for developers.

Features for Design   Typically for tool-based design, a GUI is provided that enables the specification of the appearance and behavior as well as static and dynamic content. Additionally, the tools often support the specification of interaction techniques.

As already mentioned, in order to reduce or even eliminate the need for software engineering skills, the tools provide a GUI to specify the appearance and behavior of the intended application. For this specification, D.tools, the mixed-fidelity prototyping tool from Sá, MakeIT, and OIDE are examples that visualize a state-chart diagram that can be edited by the developers: screens or speech are represented by states, whereas the user's interactions are represented by transitions. By this means, at runtime, a user interaction triggers a specified transition and leads to another state and the display of another screen. For instance, a mouse click in a screen can trigger a transition and thus a changed appearance of the application.

Most of the above-mentioned tools (e.g., D.tools and SUEDE) focus on local stand-alone applications for specific devices with static content than remote applications with dynamic content. The content of local applications with static content is known during development time. This aspect simplifies the specification because the static content (e.g., text or images) can directly be assigned to an interface element. Content, however, is often not known at specification time but instead just at runtime, such as multimedia presentations that are context-adaptively loaded and displayed. For instance, MScape dynamically displays content based on the user's outdoor location. For the specification of dynamic content, programming skills of the designers are required because behavioral rules need to be defined. In order to specify dynamic content, MScape makes use of a scripting language that enables the definition of simple logical rules.

Finally, for the design specification of modern user interfaces, a tool is expected to also support the specification of interaction techniques based on various input channels. Classically, users can interact with an application via a keyboard or mouse. In the context of the Third Paradigm, sometimes users apply more natural input channels to interact with an application (e.g., via a camera, microphone, GPS receiver, and NFC reader). In order to enable interactions via these input channels, the design tools need to support the specification of techniques of using them. MScape, OIDE, and MakeIT are examples that provide assistance for the specification and application of pervasive or ubiquitous interaction techniques. For instance, MScape enables the design specification of GPS-based locations that can be applied as user input in a mobile application. For their specification, they make use of maps that can be loaded and edited in order to input so-called location-based points of interest, for example, places with interesting sightseeing in a town.

Features for Evaluation   After the design specification, a high-fidelity prototype is automatically generated. This prototype can be used to conduct evaluations. Characteristically, a tool assists in the synchronous capturing of all user interactions together with the audiovisual content. Additionally, some tools record the appearance of the application and enable the recording of live annotations of the evaluator.

The synchronous capturing of user interactions is the key feature of an evaluation tool. For instance, all keyboard or mouse clicks of a user are saved. By this means, the analysis could reveal whether users have a typical behavior or preference when using an application, such as a website. All of the above-mentioned evaluation and analysis tools as well as all user-centered prototyping tools support this feature.

Besides the logging of user interactions, the audiovisual capture of the user while interacting with the application is a further important feature of an evaluation tool. Analysis of the logged user interactions alone often does not give a complete insight into the user's behavior and preferences. Instead, the analysis of the recorded user interactions in combination with the audiovisual content can provide valuable data for interpretation. For instance, in this way, the question can be answered as to why a subject has selected the same option several times. D.tools and Momento are examples that support the synchronized recording of user interactions and audiovisual content.

Evaluation and analysis tools sometimes also record the appearance of the application during a user evaluation. For instance, a screen recorder is used to capture the changes of the appearance, such as the different screens. Additionally, these tools also help to save live annotations of the evaluator, such as comments for the analysis.

Features for the Analysis   After the user evaluation is over, the tool is also expected to assist in the analysis of the captured data.

The main objective of the analysis is to find problems of a prototype, such as usability problems. Additionally, the analysis also helps answer questions about user behavior or preferences in different situations. Therefore, a software tool is expected to appropriately display all captured data to easily and quickly enable the analysis. A common feature is to display the audiovisual content as well as the other captured data synchronously in a timeline-based GUI. In this way, the audiovisual content is preannotated. The developers do not need to perform this by hand anymore, which is a time-consuming and annoying task. Now the developer can immediately scroll through the preannotated video or jump to intended data. D.tools and Momento apply the interactive time-line-based visualization of the logged data as well as the audiovisual content.

MoPeDT: A User-Centered Prototyping Tool for Mobile Applications

We have developed a user-centered prototyping tool called MoPeDT (Pervasive Interface Development Toolkit for Mobile Phones) that can be used to design, evaluate, and analyze mobile applications that support different interaction techniques, such as a technique that utilizes NFC. In this section, we first introduce MoPeDT's applied software architecture and modules. Then we describe the different tool features of MoPeDT.

The Architecture and Software Modules

In order to support user-centered development and evaluate mobile applications in the context of the Third Paradigm, MoPeDT employs an architecture and their software modules (see Figure 5.3). This architecture contains physical objects, mobile clients, a server, and a database as well as sensors, actuators, and evaluators.

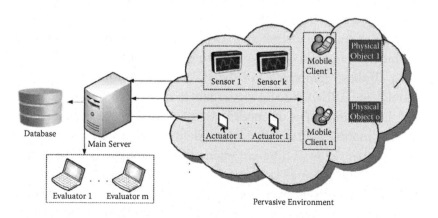

Figure 5.3   The client-server architecture that was applied for MoPeDT.

Physical objects are real objects in a pervasive environment, such as works of art in a museum. For these objects, services and content exist in the database. Users can apply their mobile phones (mobile client) in order to address the physical objects. For this addressing, users can utilize different interaction techniques that are applied via their mobile application, such as a techniques based on NFC. After having selected a physical object, the mobile client communicates with a server to get all services and contents that are stored in the database.

Sensors (e.g., a temperature sensor) are also plugged into the server in order to provide knowledge of the user's environmental context while a user is interacting with the application, such as knowledge about the temperature, lighting conditions, or loudness. Knowledge about the contextual situations can help to interpret a user behavior in the analysis of a user study. The plugged-in sensors can also be used as a further input channel for mobile clients. In this way, a particular context of a sensor (e.g., hot or cold) might cause an adaptation of the mobile phone's appearance. A further architectural component called *evaluator* is applied whenever tool-supported local or remote user studies have to be conducted. Several of these evaluators can connect to the server and register the interest in other connected components: mobile clients and sensors. Then, the evaluators can synchronously log all contexts of the selected mobile clients and sensors for later analyses. The last plug-and-play component of the architecture is the actuator, which can be used as an additional output channel for multimedia content. For instance, as described in our previous work (Leichtenstern and André 2009a), our architecture can be applied to generate a pervasive game that not only contains multiple users applying different mobile devices with different interaction techniques but also includes a public display presenting video content.

For all these components of the architecture, we provide software modules in J2SE or J2ME (Leichtenstern and André 2009b). The most interesting software module is the module for the mobile client. This module contains the entire client-server communication and the implementation of different interaction techniques based on input from the mobile phone's keyboard, touch screen, camera, microphone, NFC reader, GPS receiver, and accelerometer. These techniques can be configured by using an XML file. Additionally, the mobile client supports the usage of different screen templates based on approved

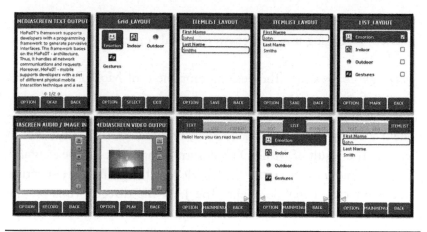

Figure 5.4 Screens generated based on the screen templates.

mobile phone guidelines from Nokia. For instance, these templates consider a consistent layout, softkey usage, and navigation style. Each screen has a heading, content, and a softkey part. In the softkey part, the left softkey is used for options, whereas the middle key is applied for confirmations and navigations, and the right softkey is utilized for negative actions (back, cancel, or exit). Additionally, each screen contains a contextual help and an option to return to the main menu. From each screen, the user can return to the previous state automatically. Thus, each screen provides a back, cancel, or exit option. Based on this feature of the mobile client software, MoPeDT also assists in compliance with approved interface guidelines in order to improve the quality of the resulting prototypes. Most of these tools do not support this feature, such as D.tools and SUEDE. Figure 5.4 shows screens that were generated based on the screen templates. For the specification of the screens (appearance) and the logic (behavior) of a mobile application, XML files are used.

The User-Centered Prototyping Tool: MoPeDT

In this section, we summarize MoPeDT's features for design, evaluation, and analysis. MoPeDT's features are mainly based on meaningful features of the just-mentioned tools, but some further features were added. By means of the aforementioned screen templates of the mobile client software, MoPeDT assists compliance with approved interface guidelines in the design. Additionally, our architecture enables a wider range of evaluations. For instance, sensors can be attached to also log user and environmental context.

The Design Component The design component of MoPeDT supports the specification of the appearance, behavior, and interaction techniques. Additionally, in contrast to D.tools and SUEDE, the component also supports the specification of static and dynamic screen content. The result of MoPeDT's design component is an executable JAR file that makes use of the previously mentioned architecture component called mobile client. The JAR file directly runs on the end-device. We successfully tested the generated high-fidelity prototypes on several Nokia phones (S40 and S60).

MoPeDT also applies the approach to support the specification of an application's appearance and behavior via a state-chart diagram view (see Figure 5.5). Since several views can provide benefits for the developer, MoPeDT not only supports a state-chart diagram view but also a tree view. For instance, MoPeDT's tree view supports a better overview when working on a specific screen, whereas the state-chart view provides a better overview of the entire specification of the appearance and behavior. When specifying a mobile application, the designer can first add several screens and edit their content. These

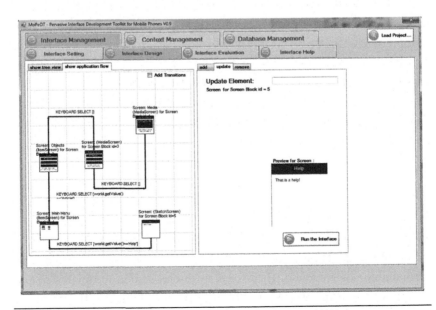

Figure 5.5 The design specification of the interface appearance and behavior.

screens are based on our mobile client software and therefore the aforementioned interface guidelines were considered. After the specification of the screens the designer can link several screens, by adding transitions. For instance, a transition can be added to a screen so that, at runtime, another screen should be loaded once a user has selected a physical object via the mobile phone's NFC reader.

MoPeDT's generated applications are remote applications that can dynamically load and display content of a database. Therefore, a special feature of MoPeDT is the support for specifying static as well as dynamic content of the prototype's appearance. Static content (e.g., images, videos, or text) can directly be assigned to the screen at the specified time. In order to enable a specification of dynamic screen content, MoPeDT also makes use of a scripting language (Leichtenstern and André 2009b). For instance, if a designer would like to specify that all stored services in the database of a selected physical should be displayed at runtime, the scripting language can be used. The scripting languages can also be used for the speciation of context-adaptive appearance; for example, the screen content should be displayed differently whenever the user is at a certain location. The specification of the database content (pervasive environments, objects, services, and content) is a further feature of the design component (see Figure 5.6).

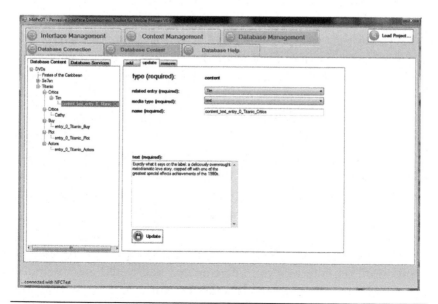

Figure 5.6   The design specification of the content in the database.

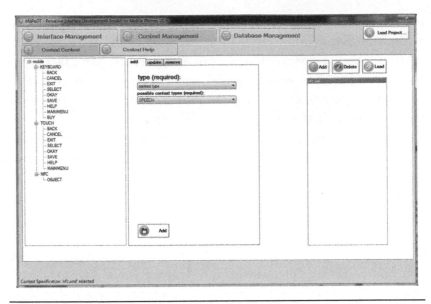

Figure 5.7   The design specification of the interaction techniques.

As previously described, MoPeDT also features the specification and employment of different interaction techniques that make use of the mobile phone's built-in hardware (e.g., the NFC reader). Figure 5.7 shows the GUI to specify these different interaction techniques. Designers can add several of the supported techniques and provide values to these interaction techniques. For instance, values of the keyboard-based interaction might be *SELECT, BACK, CANCEL*, or *BUY*. After the specification of the techniques and their values, they can be used as a transition for the specification of the prototype's behavior. For instance, the transition *KEYBOARD.BUY* can be used to switch from one screen to another once the user has selected the command *BUY* via the mobile phone's keyboard. Interactions via NFC can be specified similarly. For instance, once the developer has added the value *BUY* to the interface technique NFC, a transition can be used called *NFC.BUY*. At runtime, the transition is triggered once the user has touched an RFID tag that contains the string *BUY*.

The Evaluation Component   Once a designer has specified the mobile application, the interface can be generated and tested via an emulator or directly on the mobile phone. To run a user evaluation, the evaluation component (see Figure 5.8) supports different features. The

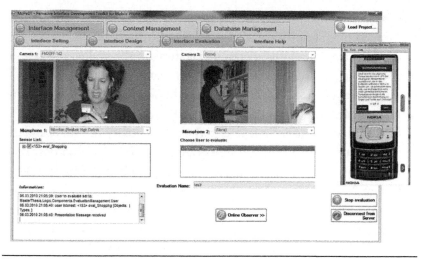

Figure 5.8   The evaluation of an evolutionary prototype.

component can synchronously capture all user interactions, audio-visual content, and the appearance of the application as well as live annotations of the evaluator. Additionally, the component also supports the capturing of the pervasive environment in terms of the user and environmental context.

The main feature of MoPeDT's evaluation component is the support to synchronously record all user interactions during a user study. This means that all events of the supported interaction techniques (e.g., keyboard-based or NFC-based) can be logged. For this feature, MoPeDT makes use of the architecture's plug-and-play component called evaluator.

Besides the recording of user interactions, a further feature of MoPeDT's evaluation component is the support for audiovisually recording the user and her environment while she is interacting with a prototype, for example, in order to complete a task.

Whenever a user study is conducted, MoPeDT displays a cloned screen view of the subject's mobile phone screen on the evaluator's computer. Thus, during the local or remote user study in a laboratory or field setting, the evaluator can always trace all interactions of the subject and the subject's current screen view. This cloned mobile phone screen is also used in order to take screen shots at certain points in time during the user study. For instance, MoPeDT captures screen shots of the cloned screen view whenever a new appearance is displayed. After the study, the captured screen shots can help analyze the logged interactions, contexts, and the captured videos.

In addition to the earlier mentioned features of the evaluation component, live annotations are a further support that is provided by MoPeDT. During the evaluation, evaluators can use this feature in order to log comments and observations or describe the executed task in more detail.

Finally, since the user context (e.g., the user's activity and state) and the environmental context (e.g., the temperature and lighting conditions) can also give valuable information for later analyses, their recording is another feature of MoPeDT's evaluation component. In contrast to most other user-centered prototyping or evaluation tools, our architecture enables the add-on and application of sensor components that enable the determination and logging of user and environmental contexts.

The Analysis Component   In the final step of user-centered prototyping, the developer can analyze the user evaluation. In order to simplify this task, the analysis component provides a GUI (see Figure 5.9) that displays all recorded data as well as a feature to export the captured data to run statistical analyses.

MoPeDT's analysis component provides the timeline-based visualization of the recorded data in order to navigate through them and interact with them (e.g., to find usability problems or user preferences).

Figure 5.9   The analysis of an evolutionary prototype.

We extended ANVIL (Kipp 2001) in order to develop MoPeDT's analysis component. ANVIL supports the display of audiovisual content as well as the visualization and modification of annotations at various freely definable timeline-based tracks. The extended version of ANVIL automatically synchronizes and annotates the captured audiovisual content with the recorded user interactions and contexts. Now, the interface developer can scroll through the preannotated video or jump to intended data that are displayed in the tracks in order to investigate the user's behavior in different contextual situations.

Since the determination of significant results is often an important analysis task, a further feature of MoPeDT's analysis component is to support statistical analyses. MoPeDT supports the export of the annotated data in different formats of statistic tools (e.g., SPSS) in order to investigate the probability of occurrence of an intended context or behavior. Thus, the subject's errors, their number, or the subject's required time to complete a task can be statistically analyzed with the assistance of MoPeDT.

The Evaluation of MoPeDT

In an evaluation of MoPeDT (Leichtenstern and André 2010), we wanted to shed light on two aspects. First, we wanted to know whether the developer's efficiency and effectiveness as well as satisfaction can be increased when using a support tool such as MoPeDT. Thus, we investigated the factors usability (ISO 9241 part 11), efficiency, effectiveness, and satisfaction. Additionally, we wanted to know the benefits and problems of using a UCP tool for user-centered prototyping. For this evaluation, we defined an experimental setting, conducted a user study, and analyzed the results.

The Experimental Setting

In our experiment we first formulated three hypotheses; then, we defined independent and dependent variables as well as the used evaluation techniques. We defined the following three hypotheses:

- H1: MoPeDT improves the developer's efficiency in quickly developing an interface prototype.

- H2: MoPeDT improves the developer's effectiveness in developing a highly user-friendly interface prototype.
- H3: MoPeDT improves the developer's satisfaction.

To investigate our hypotheses, we defined the used platform as an independent variable with the following two levels. In the first level, the developers applied a user-centered prototyping tool (MoPeDT) for the user-centered prototyping of a mobile application, whereas in the second level, the developers used a traditional approach of an IDE with an emulator (e.g., Eclipse or Netbeans) to design, evaluate, and analyze the mobile application in a single iteration. We applied different evaluation techniques in order to measure our dependent variables: efficiency, effectiveness, and satisfaction. We used a questionnaire to acquire subjective data, whereas protocol recordings and a guideline review were utilized in order to collect objective data. Later, the aggregated data were compared in order to determine the correctness of our hypotheses.

The protocol recording was applied in order to gather objective data for the investigation of hypothesis H1 (Efficiency). The subjects used the protocols for the documentation of the required time to complete different steps while designing, evaluating, and analyzing the prototypes. Moreover, problems that arose had to be noted. To gather objective data for hypothesis H2 (Effectiveness), we conducted a guideline review and investigated the resulting prototypes for both levels. An independent usability expert who was not involved in the development or evaluation of MoPeDT used the generated high-fidelity prototypes and investigated their robustness and completeness based on the description of the application as well as their violation against the mentioned guidelines based on Nokia's Design and User Experience Library.

In our posttask questionnaire, we asked the subjects to rate statements about the prototype's design, evaluation, and analysis for both levels: with MoPeDT and with the traditional approach. The statements addressed the efficiency (E), effectiveness (Eff), satisfaction (S), learnability (L), transparency (T), and their satisfaction with the interface developer's user control (C).

We used a within-subjects design and, therefore, all our 20 subjects participated in both levels of the experiment. To prevent any position effects, 10 subjects started with the design, evaluation, and analysis of the prototype based on the traditional application of the user-centered design and afterward used MoPeDT, whereas the other 10 students used MoPeDT first.

Conducting the Experiment

In both levels, that is, user-centered prototyping with and without MoPeDT, the same pervasive shopping assistant for mobile phones had to be (1) designed, (2) evaluated, and (3) later analyzed with end users of the application. This pervasive shopping assistant helps users receive information about articles in a shopping store (e.g., about the ingredients of articles). (1) To maintain comparability, the subjects received a detailed description about the intended prototype. For example, the types of screens and content to display were predefined. Moreover, the description also contained the requirement to implement prototypes that enable a keyboard and an interaction based on NFC. The subjects were also instructed to implement a logging mechanism for the prototype that was generated with the traditional approach in order to enable a recording of the user interactions in the evaluation phase. (2) For the evaluation of their generated prototypes, the subjects were instructed to audiovisually capture three end users while they were interacting with the two generated prototypes. The task for this evaluation was also predefined in order to enable a comparison between the captured evaluations of the two settings. (3) After the user evaluation, the subjects had to analyze their captured audiovisual content and logged user interactions in order to find usability problems of the two prototypes, such as wording problems. Figure 5.10 shows two prototypes that were designed, evaluated, and analyzed with one subject on the levels with and without MoPeDT.

The subjects of our studies were students of our three-month course "Usability Engineering." Before we ran the study, at the last third of the course, we conducted tutorials within the course and taught all subjects how to use MoPeDT for the interface's design, evaluation, and analysis, and how to implement and evaluate mobile phone prototypes using Eclipse with EclipseME and Netbeans with the Mobility Pack. During this training period and the user study, we did not tell the subjects that MoPeDT is a software tool that was developed at our

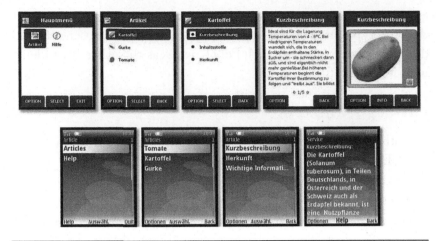

Figure 5.10   Prototype generated with (first row) MoPeDT and with (second row) the IDE.

laboratory. In addition to the software skills, we comprehensively taught all subjects about usability in general, user-centered design, mobile phone usability, and the mobile phone guidelines that the usability expert also used for the guideline review. We reminded the subjects to apply these guidelines for both levels when designing, evaluating, and analyzing the prototypes in our user study. During the design, evaluation, and analysis of the user study, the subjects were ordered to fill in the protocols for all completed steps.

Results and Discussion

Twenty computer science students (16 male and 4 female students) participated in our one-month user study. The subjects were aged between 22 and 29 (M = 24.15, SD = 1.90). Most of the participants rated themselves as medium skilled in object-oriented programming languages (e.g., Java and C++) and mobile phone programming (e.g., J2ME) as well as medium skilled in usability engineering in general and mobile usability engineering. In the following, we describe the results of our user study and discuss their consequences for our three hypotheses: the efficiency, effectiveness, and satisfaction when using the all-in-one solution MoPeDT instead of an IDE.

Efficiency   The analysis of our data proved our assumption about increased efficiency when using MoPeDT. When analyzing the protocols based on a two-sided dependent t-test, on average, the required design, evaluation, and analysis time in minutes with traditional approaches, for example, Eclipse (M = 816.60, SD = 318.81), was significantly higher than when using MoPeDT (M = 266.65, SD = 208.14) [$t(19)$ = 9.2, p < 0.001]. When not using MoPeDT, the network and GUI programming required much more time in the prototype design phase. In the evaluation and analysis phase, the annotation and analysis of the captured videos decelerated the user-centered design process when not using MoPeDT.

These objective data were also reflected when analyzing our subjective questionnaire data. Based on our rating scale from one to five, on average, the subjects agreed with the statement about their increased efficiency in (running the user-centered design process) when using MoPeDT, which was significantly higher than when using the traditional approach [$t(19)$ = 4.68, p < 0.001]. The efficiency when using MoPeDT for evaluation and analysis was also seen as higher, but not significantly. The subjects also found that MoPeDT makes the whole user-centered design process more efficient compared to the traditional approach. Additionally, the time gain with MoPeDT was significantly more highly rated than the time gain when using the traditional approach [$t(19)$ = 7.37, p < 0.001]. The qualitative feedback of the questionnaire substantiates the results. Most subjects found the tool usage "quick and easy" and saw a benefit in "the very quick prototyping and evaluation of applications." The results of the protocol recording and the questionnaire prove our first hypothesis that MoPeDT can increase the developer's efficiency compared to a traditional approach.

Effectiveness The analysis of the subjective data revealed similarly rated results for the quality of prototypes that were generated with an IDE compared to the prototypes that were generated with MoPeDT. The analysis of the qualitative data shed light on the participants' rating. While most of them highlighted the prototypes that were generated with MoPeDT as "beautiful" and "follow design guidelines," they also pointed out the limitation caused by the screen templates. For instance, one subject had this feedback: "I could not individually design the application because I had to conform to the prefabricated patterns." This finding was supported by the ratings regarding the scope of the supported screen templates. Nevertheless, based on a two-sided dependent *t*-test, the results of our guideline review showed a highly significant difference between the interfaces that were designed, evaluated, and analyzed with MoPeDT compared to the generated interfaces with the traditional approach. Interfaces generated with MoPeDT had, on average, less violations against the 22 guidelines (M = 0.85, SD = 0.93) than the interfaces that were developed with traditional approaches (M = 4.35, SD = 2.52) [$t(19)$ = 5.48, $p < 0.001$]. Most often, the interfaces developed with traditional approaches did not consider a consistent usage of the softkeys (17 of 20 subjects) and did not use icons and text for important information (17 of 20 subjects). Based on our objective data, the second hypothesis about the increased effectiveness of the developers at generating more user-friendly interfaces is proved by the fact that the resulted prototypes of MoPeDT provided better usability based on the guideline review than the prototypes generated with traditional approaches. The subjects, however, would like to have a wider range of action when designing the layout with MoPeDT.

Satisfaction The overall ratings on satisfaction are illustrated in Figure 5.11. In general, the subjects felt significantly constrained in the actions available to them when using MoPeDT compared to the range of actions supported by the IDE [$t(19)$ = 3.21, $p < 0.01$]. The limited user control is mainly responsible for the fact that the subjects were significantly more satisfied when using the IDE for the design of

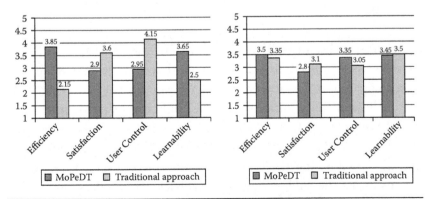

Figure 5.11 Results of the questionnaires for the (left) Design and (right) Evaluation and Analysis Component.

a prototype compared to MoPeDT [$t(19) = 2.41, p < 0.05$]. Besides the limited user control, transparency was also pointed out as a problem with MoPeDT. The interface designers want to see what is going on in the background. Despite these overall negative results in terms of satisfaction, the subjects did find some benefits with respect to learnability. On average, the ease of learnability was significantly higher rated for the design with MoPeDT than the traditional approach [$t(19) = 3.61, p < 0.01$], whereas the ease of learnability for the evaluation and analysis with MoPeDT was similarly rated as when not using MoPeDT. The subjects required fewer skills when designing or evaluating and analyzing prototypes with MoPeDT compared to the traditional approach.

In our questionnaire, we also asked the subjects about their overall acceptance of MoPeDT. Thus, we asked them to decide about their preferred level for the design as well as the evaluation and analysis. For the design, five subjects chose the IDE, whereas seven selected MoPeDT; eight subjects judged both approaches to be useful, which is quite similar to the results of the preferred evaluation and analysis approach. Four subjects favored not using a tool for evaluating and analyzing prototypes, whereas 11 subjects preferred MoPeDT; five saw benefits of using both approaches. Thus, despite the negative satisfaction with MoPeDT, the subjects preferred MoPeDT for the design, evaluation, and analysis compared to the traditional approach.

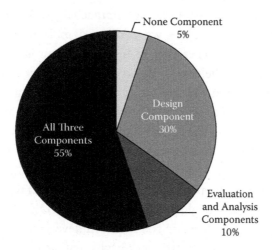

Figure 5.12   Subjects' preference for a component of MoPeDT.

We also asked about the subject's preferred components of MoPeDT (see Figure 5.12): one subject did not like a single component of MoPeDT, six subjects only liked the design component, and two subjects only liked the evaluation and analysis component. Eleven subjects liked all components of MoPeDT: the components to design, evaluate, and analyze a prototype. The qualitative feedback reveals that the participants favored the approach of having an all-in-one solution. For instance, a subject mentioned the benefit of being able to "handle everything in a single program: the database, the design, evaluation, and analysis." Another subject mentioned that "only the combination of all components meaningfully supports iterative prototyping," which is similar to the statement that quick and easy prototyping can be improved by "the close interleaving of the three components" and "the all-in-one approach" that prevents "the use of several programs."

In conclusion, we could not prove the hypothesis about the increased interface developer's satisfaction when using MoPeDT because of some problems due to the inadequate user control and transparency.

Conclusion

In this chapter, we covered different types of software support for the development of mobile applications in the context of the Third Paradigm. After the introduction of traditional approaches—IDEs with middleware and software modules—we addressed tools that assist in the design and automatic generation of high-fidelity prototypes. In order to assist all stages of the user-centered prototyping, we also considered tools for the evaluation and analysis of the generated high-fidelity prototypes. Ultimately, we introduced UCP tools as all-in-one solutions for interface developers. In particular, we introduced our UCP tool called MoPeDT that assists the user-centered prototyping of mobile applications in all phases. To assess whether developers should use a tool or traditional software support, we conducted a comparative study with our tool and the traditional approach of using an IDE.

A tool support can decrease the required development time as well as usability engineering and the required software engineering skills. Additionally, it can improve the quality of a prototype by automatically incorporating approved interface guidelines. Thus, the efficiency and effectiveness of a developer can be improved when applying a tool support. Our results also indicate that our subjects mostly accept UCP tools such as MoPeDT for all steps of user-centered prototyping. Typical problems of tool support are the limited user control and transparency. A tool can hardly provide the same range of action compared to traditional approaches of using an IDE.

Overall, tool support is strongly recommended for low- or medium-skilled developers in software or usability engineering. Different high-fidelity prototypes with wide-ranging functionalities can quickly and easily be generated and evaluated in the early stages of the development process, which is hardly possible with traditional approaches. In contrast to that, whenever the final mobile application with individualized features and layout should be implemented, IDEs with emulators and the support of software modules and middleware seem to be more feasible because they provide a wider range of actions and better transparency for developers.

Acknowledgment

This research is partly sponsored by OC-Trust (FOR 1085) of the German research foundation (DFG).

References

Andreasen, Morten, Henrik Nielsen, Simon Schröder, and Jan Stage. "What happened to remote usability testing?: An empirical study of three methods." *Proceedings of the SIGCHI Conference on Human Factors in Computing Systems.* ACM, New York 2007. 1405–1414.

Arroyo, Ernesto, Ted Selker, and Willy Wei. Usability tool for analysis of web designs using mouse tracks. *Extended abstracts on Human Factors in Computing Systems.* ACM, 2006. 484–489.

Ballagas, R., J. Borchers, M. Rohs, and J. G. Sheridan. "The smart phone: A ubiquitous input device." *Pervasive Computing,* 5(1), 2006: 70–77.

Bateman, Scott, Carl Gutwin, Nathaniel Osgood, and Gordon McCalla. "Interactive usability instrumentation." *Proceedings of the 1st ACM SIGCHI Symposium on Engineering Interactive Computing Systems.* ACM, 2009. 45–54.

Benford, Steve et al. "Savannah: Designing a location-based game simulating lion behaviour." *Proceedings of the 2004 International Conference on Advances in Computer Entertainment Technology.* ACM, 2004.

Broll, Gregor, Susanne Keck, Paul Holleis, and Andreas Butz. "Improving the accessibility of NFC/RFID-based mobile interaction through learnability and guidance." *Proceedings of the 11th International Conference on Human-Computer Interaction with Mobile Devices and Services.* ACM, New York 2009.

Carter, Scott, Jennifer Mankoff, and Jeffrey Heer. "Momento: Support for situated ubicomp experimentation." *Proceedings of the SIGCHI Conference on Human Factors in Computing Systems.* ACM, New York 2007. 125–134.

Chesta, Cristina, Fabio Paternò, and Carmen Santoro. "Methods and tools for designing and developing usable multi-platform interactive applications." *PsychNology Journal* 2004: 123–139.

Davis, A. M., H. Bersoff, and E. R. Comer. "A strategy for comparing alternative software development life cycle models." *IEEE Trans. Software Engineering,* 1988: 1453–1461.

Davis, Alan. "Operational prototyping: A new development approach." *IEEE Software,* 1992: 70–78.

de Sá, Marco, Luis Carrico, Luis Duarte, and Tiago Reis. "A mixed-fidelity prototyping tool for mobile devices." *Proceedings of the Working Conference on Advanced Visual Interfaces.* ACM, 2008. 225–232.

Dey, Anind, Raffay Hamid, Chris Beckmann, Ian Li, and Daniel Hsu. "a CAPpella: Programming by demonstration of context-aware applications." *Proceedings of the SIGCHI Conference on Human Factors in Computing Systems.* ACM, 2004. 33–40.

Dow, Steven, Jaemin Lee, Christopher Oezbek, Blair MacIntyre, Jay Bolter, and Maribeth Gandy. "Exploring spatial narratives and mixed reality experiences in Oakland Cemetery." *Proceedings of the 2005 ACM SIGCHI International Conference on Advances in Computer Entertainment Technology.* ACM, 2005. 51–60.

Froehlich, Jon, Mike Y. Chen, Sunny Consolvo, Beverly Harrison, and James A. Landay. "MyExperience: A system for in situ tracing and capturing of user feedback on mobile phones." *Proceedings of the 5th International Conference on Mobile Systems, Applications and Services.* ACM, 2007. 57–70.

Greenhalgh, Chris. "EQUIP: A software platform for distributed interactive systems." *15th Annual Symposium User Interface Software and Technology.* 2002: ACM, 2002.

Hartmann, Björn, and Scott Klemmer. "Reflective physical prototyping." *Proceedings of UIST 2006 Symposium on User Interface Software and Technology.* ACM, 2006. 299–308.

Holleis, Pau, and Albrecht Schmidt. "MakeIt: Integrate user interaction times in the design process of mobile applications." *Proceedings of Pervasive.* Springer, 2008. 56–74.

Holmquist, Lars Erik. "Prototyping: Generating ideas or cargo cult designs?" 2005: 48–54.

Houde, Stephanie, and Charles Hill. "What do prototypes prototype?" In *Handbook of Human-Computer Interaction,* by Martin Helander, Thomas Landauer and Prasad Prabhu. Elsevier Science, Amsterdam, 1997.

Hull, Richard, Ben Clayton, and Tom Melamed. "Rapid authoring of medi-ascapes." *Proceedings of 6th International Conference of Ubiquitous Computing.* Springer, Heidelberg, 2004. 125–142.

Kangas, Eeva, and Timo Kinnunen. "Applying user-centered design to mobile application development." *Communications of the ACM,* 48(7), 2005, 55–59.

Kindberg, Tim et al. "People, places, things: Web presence for the real world." *Proceedings WMCSA 2000.* IEEE, 2000. 365–376.

Kipp, Michael. "Anvil—A generic annotation tool for multimodal dialogue." *Proceedings of the 7th European Conference on Speech Communication and Technology.* 2001. 1367–1370.

Klemmer, Scott, Anoop Sinha, Jack Chen, James Landay, Nadeem Aboobaker, and Annie Wang. "Suede: A wizard of Oz prototyping tool for speech user interfaces." *Proceedings of the 13th Annual ACM Symposium on User Interface Software and Technology.* ACM, 2000. 1–10.

Leichtenstern, Karin, and Elisabeth André. "MoPeDT—Features and evalua-tion of a user-centred." *Proceedings of the 1st ACM SIGCHI Symposium on Engineering Interactive Computing Systems.* ACM, 2010.

Leichtenstern, Karin, and Elisabeth André. "Studying multi-user settings for pervasive games." *Proceedings of the 11th International Conference on Human-Computer Interaction with Mobile Devices and Services.* ACM, 2009a.

Leichtenstern, Karin, and Elisabeth André. "The assisted user-centred genera-tion and evaluation of pervasive interfaces." *Proceedings of the European Conference on Ambient Intelligence.* Springer, 2009b. 245–255.

Li, Yang, Jason Hong, and James Landay. "Topiary: A tool for prototyping loca-tion-enhanced applications." *Proceedings of the 17th Annual ACM Sympo-sium on User Interface Software and Technology.* ACM, 2004. 217–226.

Macleod, Miles, and Ralph Rengger. "The development of DRUM: A soft-ware tool for video-assisted usability evaluation." *Proceedings of HCI.* Cambridge University Press, 1993. 293–309.

Mao, Ji-Ye, Karel Vredenburg, Paul W. Smith, and Tom Carey. "The state of user-centered design practice." *Communication of the ACM,* 48(3) 2005: 105–109.

Mattern, Friedemann, and Christian Floerkemeier. "Vom Internet der com-puter zum internet der dinge." *Informatik Spektrum,* 33(2), 107–121, April 2010.

McGee-Lennon, Marilyn Rose, Andrew Ramsay, David McGookin, and Philip Gray. "User evaluation of OIDE: A rapid prototyping plat-form for multimodal interaction." *Proceedings of the 1st ACM SIGCHI Symposium on Engineering Interactive Computing Systems.* ACM, 2009. 237–242.

Myers, Brad A. "User interface software tools." *ACM Transactions on Computer–Human Interaction,* 2(1), 1995: 64–103.

Rukzio, Enrico, Gregor Broll, Karin Leichtenstern, and Albrecht Schmidt. "Mobile interaction with the real world: An evaluation and comparison of physical mobile interaction techniques." *Ambient Intelligence, European Conference.* Springer, 2007. 1–18.

Rukzio, Enrico, Karin Leichtenstern, Vic Callaghan, Paul Holleis, Albrecht Schmidt, and Jeannette Chin. "An experimental comparison of physical mobile interaction techniques: Touching, pointing and scanning." 87–104. 2006.

Salber, Daniel, Anind Dey, and Gregory Abowd. "The context toolkit: Aiding the development of context-enabled applications." *Proceedings of the 1999 Conference on Human Factors in Computing Systems,* Pittsburg, PA. ACM, New York, 1999.

Weiser, Mark. The computer for the 21st century. *Scientific American,* 1991.

6

UNIVERSITY OF THINGS

Toward the Pervasive University

IRENE LUQUE RUIZ, PILAR CASTRO
GARRIDO, GUILLERMO MATAS MIRAZ,
FRANCISCO BORREGO JARABA, AND
MIGUEL ÁNGEL GÓMEZ-NIETO

Contents

Near Field Communication: The Contactless Revolution

In the computing environment, the Internet of Things (IoT) refers to a network of objects that are able to interconnect among themselves for exchanging information (Weiser 1991). The idea is quite simple, but not its deployment. It is about introducing small tags (RFID labels) into everyday familiar objects so they can be identified and handled by a computer in a virtual world as they are handled by people in the real world.

This is the meaning of the term *Ambient Intelligence* (AmI), which describes an environment where people would be surrounded and aided by smart and intuitive interfaces embedded in everyday objects and communicate with them, creating an electronic environment that would recognize and respond to the presence of individuals immersed in it in an invisible and forward-looking way (Aarts, Harwig, and Schuurmans 2001).

Through natural interactions, the user can access a set of services that are provided by the objects that surround them, defining an interaction scenario. These services are adapted to the context and preferences of the interacting user, so that they could help with the user's daily activities.

The central feature of the envisioned scenarios is that people are at the forefront of the information society. This vision of people benefiting from services and applications while being supported by new technologies in the background and intelligent user interfaces was essential to the Information Society Technologies Advisory Group (ISTAG) notion of AmI in the first place (ISTAG 2001).

Some features of AmI according to ISTAG are (a) ease of human contact, (b) helping to create knowledge and skills for work, (c) inspiring trust and confidence, and (d) usable by everybody.

The paradigm of ubiquitous or pervasive computation (Weiser 1993) allows the development of active and mobile environments allowing access to a huge amount of information and its processing, independent of the user's location, thanks to the use of computational elements intercommunicated by wireless networks, radio, GSM, etc. These elements are embedded in furniture, belongings, posters, machines, books, etc.

One of the technologies that allow us access to the digital information available in physical objects in the real world is RFID technology. The development of radio frequency identification (RFID) and its low cost has displaced the use of barcodes as a way to identify objects because RFID allows the objects to be univocally identified by tags. Besides, those tags allow bidirectional communication between the tag and the reader (Roberts 2006) in an operating range of about 30 cm to 100 cm, depending on the characteristics of the signal (HF, LF) and the size of the antenna.

Through this technology, we make use of two devices: RFID tags that will be used for storing the digital information about the physical objects and an RFID reader that, using radio waves, tests the environment looking for RFID tags to obtain information from. An RFID tag is a microchip that contains an antenna and that is placed in a physical object in order to equip it with interactive capabilities, answering the radio waves broadcasted by RFID readers.

When physical objects have tags associated with them, the user performs an interaction process for obtaining information about the physical object from which it needs the information. In Rukzio 2007, a first analysis and the classification of physical interactions with mobile devices are presented, describing some experiences, guides, and methods for easing the use of those physical interactions as well as some applications, and an analysis in privacy terms. In this paper, a set of paradigms based on three types of interactions is proposed: pointing, scanning, and touching. Scanning is useful when the user wants to discover the available services in the area around him. In this way, the services encoded in the tag are detected and shown in the user's mobile device in order to be read by any application. Pointing is used when the user can see the tag and wants to access the information stored in it. For that, the natural way for the user to interact is to point the tag to be read with his device. Lastly, touching implies physically touching the tag with the reader to select it.

Related to this last paradigm, near field technology (NFC) (ECMA 2004) has appeared. NFC is an evolution of RFID that includes, besides the capacity to communicate with radio, network and smart cards (Smart Card Alliance 2002) functions. NFC operates in a very short range (less than 10 cm), which implies that the reader and the tag must be almost touching. NFC is a technology supported by different mobile devices (phones, PDAs, etc.) that includes a secure element (Smart Card) and an NFC modem and that allows secure and bidirectional communication with another NFC device reader or tag-augmented object. Those characteristics make NFC specially suited for paying and ticketing systems, because it includes the security characteristics of the secure element (Smart Card). With this element, transactions must be made within a short range, avoiding hacking issues with the data transmitted (Newitz 2006) with devices such as RFID Cloner.

NFC will generate a new technology revolution in the forthcoming years: the contactless revolution, allowing offering ubiquitous services to users with the need for information at different times and places, supporting what has been called the IoT.

One of the main environments where those ubiquitous services are being deployed are U-cities, or ubiquitous cities, cities where everything is interconnected and where information is available ad hoc and there are no communication limits between people and computers, showing that NFC is a technology that could be used for something more than just payments and that could be combined with other technologies in order to build systems as complex as ubiquitous cities.

South Korea is leading the implementation of ubiquitous cities with projects in Seoul, Eunpyeong New Town, Songdo, Pusan, Cheju Island, and Pangyo, among others. The New Songdo city project (Songdo project 2010), the biggest one implemented in South Korea, aims to develop a ubiquitous city where all the information systems are interconnected and computers are integrated in buildings, streets, and offices.

Other world initiatives of ubiquitous cities include Dubai, Osaka, London, Docklands, Helsinki virtual village, Osterad district in Copenhagen, and technological parks such as One North of Singapore and Sabians in Brazil. It can be said that the ubiquitous computation will guide urban design.

In Europe, the city of Oulu is leading the use of NFC for the building of a ubiquitous city. The "city of Oulu-Smart touch project" (Smartouch 2008) wants to get NFC closer to all the citizens, making them responsible for the deployment and development that its use entails. Different services have been designed and implemented in order to introduce this technology in the daily life, infrastructure, and public services of the city. Among them, we can highlight (VTT 2009): (a) "hot in the city," which, through the social networks, allow anyone to get a user's location acknowledged by his friends by just bringing his device close to the tags found in the place the user is; (b) "bus ticketing service," used for urban transport payment; (c) a meal delivery service for the physically challenged; (d) "Smart Parking," a system for vehicle parking and control; (e) "Smart Theatre," information and selling system for cultural events; (f) "Information Tags in City Environment," an information system placed in buses and several city places; (g) "Future Shop Concept," a payment system for retail stores: (h) "NFC-Enabled Blood Glucose Measurement," a system that allows one to measure glucose levels in blood and to send the results in order to provide the patients the proper medication in real time; or (i) introducing NFC in schools, offering games, services for absence control and student follow-up, information, or teaching.

As can be observed, in the initiatives under way and prototypes being built around the world, the contactless revolution and the IoT guided by the NFC technology are going to include not only commercial applications for identification, payment, or ticketing, but any kind of services in any kind of environment and scenario.

The education environment is one of them, emerging in recent years the idea of the University of Things (UoT) based on the application of the IoT concept to the university environment (Matas, Luque, and Gómez-Nieto 2009a). The basic idea of this paradigm is to bring the university close to students and the workers, as well as to society in general, offering university services so that they will be available any time, any place.

Currently, people must look for and adapt to these services, while the aim is to get the services, and therefore the software, to become as mobile as the users, taking advantage of the constant change of context where they are used, resulting in active environments where computers interact among them and with the user in a smart and noninvasive way.

In a pervasive university, similar to the one proposed in this chapter, all the elements that people find in their environment are potential smart objects, thanks to the incorporation of tags in them. Thus, for example, a poster of the ones distributed across the campus (or the city) could be a registering point, as well as being an information point, so when a student moves his mobile device close to the poster, an interaction, where one of the services is the possibility of registering on the desired or announced subject, begins.

UoT is an idea that has been exploited by research groups and companies for many years under different perspectives. Across the chapter, different ways of adapting and improving the services offered by universities are described, so a ubiquitous university could be considered a closer and more accessible university where the community and society have information on demand irrespective of their location or time of the day.

Some services, for instance, the informational ones, are easy to adapt: taking into account the storage capacity of the tags, these services could be made available just by adding a tag to a poster, sign, or object used to provide information. Other more complex services, such as those that require some "intelligence," such as database access, computational resources, media resources, security, etc., could be more difficult to implement.

The university being a teaching entity, there is a need for ubiquitous applications that help the student improve his academic performance, such as ubiquitous games to motivate students, help in the library to find the material, tutorial management, etc., apart from secure payment and ticketing systems (transport, vending machines, canteens, reprography, etc.), location, and surfing (city and campus), etc.

NFC as the Appropriate Technology for the Development of the Pervasive University

Currently, universities offer many varied services to their users, for example, administrative services, such as registrations, validations, certificates, scholarships and contracts management, academic attendance and compliance control, payment management, identification and access, information, etc.; teaching services such as tutorials, examination information, schedules, classes, study abroad management, virtual teaching, etc.; services oriented to students such as book lending, tutorials and teaching guides, university public transport, shops and canteen, sports facilities, etc.

However, and for different reasons, most of them could be poor or hard to access. Therefore, with the aim of getting the universities closer to the new management and teaching models implemented in Europe during recent years (Clausen 2005), and with the purpose of achieving the maximum quality possible, we look for alternative ways to offer those services. If we factor in the advance of the knowledge-based society and the promotion and importance of the use of new technologies, NFC technology appears to be the most appropriate tool to implement the University of the Future.

To this end, different authors have proposed the use of ubiquitous computation and touch paradigms (Päivi et al. 2007), which work with NFC for the creation of an intelligent ambience in the university environment.

Nava et al. (2009) study the development of new interfaces in order to improve certain processes and to help in the daily student activities. Specifically, their proposal makes use of RFID, NFC, and sensor (infrared) technologies to facilitate the exposition process in class. Through a set of sensors, RFID devices, and NFC-enabled devices, users can automatically start their presentations (PowerPoint), as well as move forward or backward through the slides just by interacting with their device and the RFID tags or by mimicking. The authors of this paper evaluated the system based on the experiences and comments of a group of teacher-training students that used it, who accepted the system enthusiastically due to its ease of use, its applicability, and previous knowledge. The evaluation also mentioned favorably the introduction of these types of technologies in the teaching process.

Lopez de Ipiña (2007) proposes possible solutions based on RFID and NFC supported (or helped sometimes) by Bluetooth (for allowing the communication among devices) that could be used in the university environment. Among them, the following must be emphasized: (a) *Touch2Open*, an NFC-aware service to enable a user to open his office door by simply bringing a mobile device close to an RFID tag on the door; (b) *Touch2Launch*, which allows one to start and make use of services and applications just by touching an augmented object (with a RFID tag); or (c) *Touch2Print*, a service that enable users to print files just by bringing an NFC-enabled mobile device close to a PC connected to a printer.

The proposal developed by the authors in their paper force communication with an intermediate Bluetooth service for getting to each service. It makes sense for *Touch2Lauch*, but the latest improvements in NFC technology leave out this intermediate step. As an example, currently there are locks, printers, and even technical and scientific material with NFC technology already incorporated and that are able to interact directly with the NFC devices without needing the intermediate Bluetooth connection, simplifying the action.

González et al. (2008a) present other alternatives for the use of NFC in spaces dedicated to teaching. Among the proposed prototypes, we single out: (a) *touching note*, which makes use of NFC tags with a text note information stored, which are placed in the teacher's office door in order to give relevant information when the teacher is not in; (b) *touching cabinet,* in which an NFC tag is associated with a space or object so that it could provide information to the students just by touching it; (c) *campus recommender,* which basically is an application that gives guidance to new students and people who do not know much about the campus and need information about the places of interest or current location; and (d) interactive exposition panels, where the mobile phone is used for touching a surface and interacting with the information. In order to implement this last prototype, the NFC Interactive Panel uses a phone with Bluetooth and NFC capabilities. NFC provides the "touch" functions and Bluetooth the communication ones. NFC tags are set in the surface and via J2ME and JSR-172, the panel connects to a server that displays the action (in http://www.youtube.com/user/gusramir the authors show a video of the application). In other papers (Gonzalez et al. 2008b; Ramírez et al. 2008), the authors propose the concept of Touching Learning Environment as part of the evolution of M-learning and Ubiquitous Computing, including touching technologies such as NFC in mobile devices.

Lim et al. (2009) introduce in their paper the use of new technologies in the role-play, what the authors believe to be a powerful teaching tool. Within the eCIRCUS project, they have designed a framework for a technology-enhanced role-play with the aim of educating youngsters in intercultural empathy. They describe in their paper the different components of their role-play technology by means of a prototype implementation of this technology, and the pervasive game ORIENT showcase that uses NFC-enabled devices such as the Nokia 6131 NFC among other interaction methods.

Bravo et al. (2008) make use of RFID and NFC new technologies to define a new interaction model in intelligent environments where the users' presence is taken into account so they could be offered information tailored to their preferences with the university as the preferred test scenario. Therefore, the authors have developed prototypes for accessing offices, classes, or laboratories. They have proposed the development of several services: place location, visualization, information, retrieval, etc.

Glover et al. (2005) have implemented a new virtual blackboard system. Those blackboards could be used either in classrooms with teaching, or for informational purposes, or in offices and research laboratories. Through the NFC interaction of an NFC-enabled device with different tags placed around those virtual blackboards, the content to be shown could be selected as well as the answers to problems.

Other authors such as Zender et al. (2008) propose new architectures and communication models for getting a pervasive university environment by developing two E-learning applications. The first one is intended to facilitate the participation of students in remote lectures, and the second is aimed at improving and contributing new services to the E-learning platforms, which are more and more used as a teaching resource.

Zender and Tavangarian (2009) reveal the need for universities to review and redesign their organizational structure, as well as their need to adopt new solutions and services to gear up to meet the changes that universities have been going through in the last few years; and examining and reviewing the evolution of traditional universities in "pervasive universities" at the infrastructural level. They also evaluate the service-oriented university as the next infrastructural step to the desired pervasiveness, and suggest an infrastructure model for educational institutions, review the challenges and possibilities of the model, and reveal its benefits for the University of Tomorrow.

Another author who has investigated the improvement and development of the context-awareness teaching ambiance is Schmidt (Schmidt 2006). He proposes a three-tier model for the handling of information related to the interaction context at any level. This model was implemented in the project "Learning in Process," which is intended to support a new type of learning process for workplace learning: context-steered learning. Instead of the human resource development technicians being the ones who assign courses to the employees, they leave that decision to the employees, who have to actively search for learning resources, satisfying their knowledge need. The LIP system continuously monitors the employees' working activity (i.e., their context) and infers the possible knowledge gap from it (with the help of some domain knowledge). Based on this gap, the system can recommend relevant learning resources to the employees.

UoT Projects at University of Córdoba

At the University of Cordoba, and with the aim of channeling its activities for the adoption and adaptation of the new directives set by the EHEA (AQAHE 2005), our research group has developed and implemented several systems that reveal the use and power of NFC for that purpose. These prototypes have covered different types of applications aimed at students, staff, and teachers.

Smart Posters in University Scenarios

A smart poster is basically a poster where the icons, text, or images on it are augmented by RFID tags. The tags provide information and services that could be accessed by users when they touch the tags with their NFC device. If the tags are recorded through NDEF structures making use of the NFC Forum standard (NFC Forum 2010), the NFC device does not require any specific software to understand and access the services provided by the tag (i.e., receive information, make a call, connect to a web page, etc.). For more complex services, as a customized interaction guided or not through menus, the secure element in the device is accessed. Besides, when a format is proprietary and is used to record the tags, the use of a MIDlet is needed to guide the operation.

The university environment is especially suitable for the use of smart posters, which help bring information and university services closer to society in a cheaper way. Over the last few years, different projects with different difficulty levels have been deployed in the University of Cordoba.

Information Smart Posters Those systems are intended to offer information about the university, its organization, and facilities to users. Different smart posters have been developed for departments, centers, and facilities. The core idea of these systems is to use a standard virtual design where the information useful for the user about the departments, centers, or facilities is provided in the most direct way. As shown in Figure 6.1, each icon (image) showed in the smart poster has a tag that offers information and services related to the icon associated with it.

For example, the department smart posters show information (images) of the department's staff and research groups. When a user touches the picture of a teacher, he can, among other actions:

- Get the teacher's information: name, category, office location, tutorial schedule, etc.

Figure 6.1   Interacting with University Smart Posters

- Access the teacher's personal page, news about the subjects he teaches, latest marks, etc.
- Request a meeting with a teacher during his tutorial schedule.

These smart posters (Luque and Gómez-Nieto 2009) can be spread throughout the university campus, or, as in the case of the University of Cordoba, put up in any of the centers located in a wide zone of the city of Cordoba, which implies, because of the low cost, great advantages for university students.

These advantages are clearly revealed with the faculty smart posters (Matas, Luque and Gómez-Nieto 2009c). These systems offer information about faculty such as their location, degrees, subjects, rules, schedule, whether they are teaching and general staff, etc. Besides, the systems offer shifting services with the aim of facilitating and speeding up the administrative management of the centers. Therefore, a user who brings his NFC device close to the Smart Poster icon receives a token for meeting a representative of the administrative management. This token is received and stored in the NFC device with information about the day, time, and table (if needed) of the appointment. When the user gets to the office, the token is checked by a reader placed in the installations, and information about the process is gathered by the staff who performs the task. This allow the university to gain valuable information about the administrative tasks performed, their duration, objective, workload of the staff, etc., information that could be used to evaluate and improve the services offered.

In order to perform this last task, three applications must be developed: (a) a token-dispatching service located in a back-end system with the intelligence and resources needed to manage all the requests, (b) a MIDlet in charge of communication among the users and their devices with that service, and (c) a management system that collects all the information about the process and generates reports for analysis and valuation.

Compliance and Attendance Control

The new guidelines set by the EHEA lead to a pervasive university from the teaching point of view. Teaching could be in person, virtual, and anywhere, not only in classrooms, allowing strong teacher–student collaboration and facilitating the practical experience that leads to competent teaching, rather than content teaching. The deployment of this model is a difficult task for the university from the point of the teaching organization and the use of resources and facilities, including obtaining the cooperation of students and teachers as well as checking the performance of the academic schedule and facilities used.

Traditionally, control of attendance and compliance with academic norms is done in a rudimentary and inefficient way using signing sheets for teachers, a list for students, etc.; all these procedures need manual manipulation of the information afterward, which could lead to mistakes and higher costs. In order to solve those problems, a system for controlling compliance with academic norms that uses the intrinsic characteristics of the NFC-enabled devices has been developed (Matas, Luque, and Gómez-Nieto 2009b).

The infrastructure needed for implementing the system and the hardware requirements are limited and cheap: (a) the users' NFC devices store in their secure element their location (the same information that is stored in the personal university card, traditionally used by students, teachers, and staff); (b) readers installed in facilities where there is teaching activity (classrooms, laboratories, seminaries, etc.), which are connected to

Figure 6.2   Interaction process of the controlling system for academic compliance and image of the web portal applet.

a back-end system that runs software with the intelligence needed to receive and manage the information sent, as well as handle it and transform it into useful information. The communication between the readers and the back-end system is quite simple as it is handled through an applet embedded in a web portal, as can be seen in Figure 6.2.

When a user (teacher or student) accesses his space for giving or attending a class, he touches with his NFC device the reader placed there (Figure 6.2). As can be seen in Figure 6.3, the MIDlet installed in the mobile device gets the unique identification of the user from the secure element, and it is transmitted to the back-end system together with the reader ID, so the system can track the access of people to different places (classrooms, laboratories, etc.) The back-end system accesses the database verifying the authenticity of the information received, and the result is sent back to the user. In this process (Figure 6.3), the teacher can communicate issues or updates afterward, such as the subject to be taught, classroom changes, etc., and

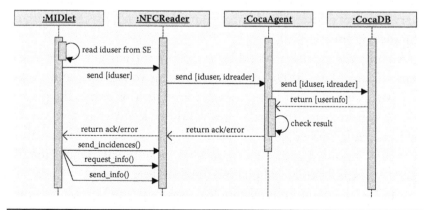

Figure 6.3   Sequence diagram of the interaction process of the academic compliance control system.

can receive ad hoc information about the students in class or request information from the system.

A web information system allows analyzing afterward the information gathered, generating reports about attendance, academic compliance, incidents, deviations with respect to the theoretical schedule, etc. This information allows evaluating either the students' behaviour or the teachers' compliance and, mainly, allowing the managers to obtain valuable information that would be used in future planning of resources and academic facilities.

Pervasive Game for Students' Motivation at University of Córdoba

Motivation is what leads us to perform activities. We are motivated when we have the will to do something and, besides, we are able to persevere in the effort required for the time needed to attain the proposed objective. For getting good academic results, students must have the "will" as well as the "skills," so there is a need for integration of these two concepts.

Student motivation is one of the challenges to overcome for the fulfillment of the EHEA directives. Ensuring student motivation and the active involvement of students in the learning process requires that it must be made attractive for them, otherwise they will not be able to summon up the necessary effort and perseverance.

A good method of motivating students is to incorporate humor into the learning process; games have also been demonstrated to be a good tool. Many games help develop certain skills and are used to carry out a set of exercises that have an educational, psychological, or stimulating function.

Among the plethora of existing games, virtual games or video games (Eow et al. 2009) are popular among youngsters, and this is why we designed a real-time strategy game. In this game, the players will be the students, the challenges would be knowledge tests on the subjects being studied, and the prizes will be credits for the academic curricula. Strategy games are games or amusements where intelligence, technical skills, and planning are vital for victory.

NFC is the appropriate technology for building pervasive games where the board depicts a real-world scenario and players interact with the objects of the real world through a common mobile device.

In order to check the utility of this technology for developing educational games, and with the aim of motivating students to learn, a strategy game was developed. In this game, a player can take on either of two available roles: seeker or pursuer. They also have to develop a set of strategies, managing the resources available, in order to get to the objectives, gain tokens or points, etc. Winning tokens or points is achieved by correctly answering the questions on the subjects the student is studying (Matas et al. 2009).

In this game, players are provided with a view of the real-world environment (game board) through a map that can be viewed in their NFC mobile device. This device must be carried by the player throughout the game and will allow him to interact with the game, acknowledge its objectives, fight other players, request information (such as help or game status), and besides, it can be used for examination preparation because to pass every objective in the game, the student must answer an academic question chosen randomly from the existing Moodle questionnaires (Moodle 2009) prepared by the teachers of the subjects the student is studying.

The development of the game can be followed by classmates, teachers, or the general public through a graphical interface that could be installed in any computer with Internet access (see Figure 6.4). This interface offers different views of the game status, depending of the level of detail expected: general game status, objective status, or a specific player's status.

The system developed is fully parameterized, what allows one to define different game sessions, with different objectives and difficulty levels, adapting it to the way the teacher evaluates, the number and type of players, the scenario, etc. This task is performed through a web

Figure 6.4 Sequences of the NFC game: objectives location, fights between players, and real time follow-up.

platform before the game starts, and allows obtaining reports before and after the game.

During the development of the game, when the players find an objective and "touch" its tag with their NFC device, they receive a question randomly chosen from the questionnaires previously prepared by the teachers of the subjects the student is studying. Those questions could be correctly or wrongly answered, and this would result in, respectively, rewards or penalties associated with the objectives. This information is displayed in real time in the graphic interface for the game follow-up. At the end of the game, the player who got to the goal first would be the winner, which would be translated into an academic prize (i.e., a number of free credits to enroll in other subjects). The winner is the player with the highest number of correct answers in the questionnaire.

Supporting Student Navigation in University Campuses

By and large, the university environment is made up of campuses. Each campus covers a large area and has several buildings: library, faculties, departments, classrooms, laboratories, sports facilities, etc. A newcomer would find it difficult to find his way around the campus. This task would become even more difficult when the campus receives students from abroad.

Traditionally, the use of maps has helped in such scenarios, but systems such as GPS and electronic maps, provided by Google (Google Maps 2010) are not appropriate, because of the nonavailability of pictures or images of the campuses. A simple and cheap solution proposed has been the use of smart posters and NFC for a guiding and location system for the students on a university campus.

This system is based on the distribution of smart posters at various locations in the university environment, made up of text and graphic information that identifies important places useful for students (secretaries, departments, buildings, zones, etc.) and that have tags associated that allow the user, through an NFC device, to access the information and surfing services.

In order to see the advantages of this system, let us think about the following situation: a student from the Hangemberg University (let us call him Hans), thanks to an Erasmus grant (European Commission 2009), is going to take up a residency in the University of Cordoba.

Figure 6.5 Smart Posters and navigation in the University Campus

On the first day, Hans goes to the Rabanales University Campus and wants to go the Computer Sciences department. This campus is on the outskirts of the city and has an area of 492.581 m² with 54 buildings.

When Hans gets to the campus's main entrance, he finds a smart poster with an image and text with information about "Departments" (see Figure 6.5). Hans "touches" this image with his NFC mobile, triggering the following actions:

a. A service associated with the tag detects that Hans's phone does not have the required MIDlet installed. The device shows Hans the option to install, and a GPRS/UMTS connection is established with the back-end system that downloads the MIDlet and installs it in the device, notifying Hans that he is able now to interact with any "intelligent" object of the University of Cordoba.

b. Hans touches again the "Departments" icon, receiving a list of the existing departments in the campus. He navigates through the list and selects the "Computer Sciences Department."

c. The MIDlet gives him different options: general information, teachers, etc., and a map (see Figure 6.5).

Hans chooses the map option, and the information required is sent to the server so he receives in his mobile device a map with the route to locate the department from the point the interaction has been made, apart from any text information needed.

The proposed solution just requires smart posters made up of a set of images, text information, and tags associated with those images offering services and information in an NFC interaction.

Tags store information about the location of the smart posters and information for identifying the location of the image (building, department, service, etc.) with which the tag is associated. When a tag is touched with a mobile device, this information is sent to a back-end system in charge of providing information about the image associated with the tag and to build a map with the route between the two places.

Navigation and location in a university campus is not different from surfing and locating points of interests in other types of open environments. Because of that, this application has been used in an interactive guide for locating points of touristic interest and surfing in the city of Cordoba. The system is named CoLoSuS (Contactless Location and Surfing System) (Borrego et al. 2010) and uses smart posters distributed in urban environments favored by tourists.

Each one of the smart posters lets tourists know the points of interest close to the location of the poster. When a tourist touches any of the images on the smart poster, the associated tag provides information and services about the place chosen, including a map that allows the user to locate it through a customized route that takes into account other points of interest recommended and points previously visited (see Figure 6.6). Besides, the system includes other functionalities to help answer questions regarding urban surfing and place location such as where am I?, how can I find it?, what is there around me?, etc.

Figure 6.6   CoLoSuS: (Contactless Location and Surfing System)

Current Status and Future of NFC for the Development of the UoT

Near Field Communication is a new technology that is being quickly deployed thanks to its ease of use, security, device availability, and easy applicability to several real-life scenarios. Its initial purpose as a secure payment system, thanks to the addition of the secure element to mobile devices, has recently evolved into the use of the SIM card for storing information in a secure way for identification and payment operations. Countless applications have been developed in recent years where NFC has proved its use for building intelligent environments. The university is one of those scenarios where NFC could be usefully deployed, offering new services and improving the existing ones, thanks to the fact that those services could be accessed anytime, anywhere just by touching a tag with a mobile device.

In this chapter, we have introduced some of the research and projects that are currently being carried out to build a pervasive university. We have presented the advantages of the use of NFC in the university environment. All of these actions have a common objective: the building of a ubiquitous university, offering administrative and teaching services to the people, in any place and at any time.

Currently, it can be said that this is the beginning of the University of Things, that there is still lots of work to do to make it a real University of the Future, where anyone with an NFC device could have full access to a set of services and information just by touching a tag or reader anywhere in the city or university campus. However, the massive production of mobiles with NFC, awaited for 2011, the use of the SIM card as a secure element, and the development of research and products would made the University of Things a reality quite soon.

Acknowledgment

This work was supported by the Ministry of Science and Innovation of Spain (MICINN) and FEDER (Project: TIN2009-07184).

Bibliography

Aarts, E., Harwig, R., and Schuurmans, M. 2001. Ambient intelligence. In *The Invisible Future: The Seamless Integration of Technology into Everyday Life*, 235–250. McGraw-Hill, New York.

AQAHE. 2005. Standards guidelines for quality assurance in the European higher education area. European Association for Quality Assurance in Higher Education. http://www.ond.vlaanderen.be/hogeronderwijs/bologna/documents/Standards-and-Guidelines-for-QA.pdf

Borrego-Jaraba, F.M., Luque Ruiz, I., and Gómez-Nieto, M.A. 2010. A pervasive solution NFC based for city touristic surfing. *Personal and Ubiquitous Computing*. Submitted for publication.

Bravo, J., Hervás, R., Chavira, G., Nava, S.W., and Villarreal, V. 2008. From implicit to touching interaction: RFID and NFC approaches. *IEEE Conference on Human System Interaction (HIS'08)*. Pages 743–748. Krakrow, Poland. May 2008.

Castro, P., Matas, G., Luque, I., and Gómez-Nieto, M.A. 2009. Seek-it and Touch-it: A Pervasive Game using Mobile Phones and Near Field Communication Technology for Players Interaction with Smart Scenarios. Mobile Networks and Applications. Submitted for publication.

Clausen, T. 2005. Undergraduate engineering education challenged by the Bologna declaration. *IEEE Transactions on Education*, 48(2), 213–215.

ECMA. 2004. Near Field Communication white paper. http://www.ecma-international.org/activities/Communications/2004tg19-001.pdf (accessed February 18, 2010).

Eow, Y.L., Ali, W.Z.B.W., Mahmud, R.bt., and Baki, R. 2009. Form one students' engagement with computer games and its effect on their academic achievement in a Malaysian secondary school. *Computers & Education*, 53, 1082–1091.

European Commission. 2009. The Erasmus programme. http://ec.europa.eu/education/lifelong-learning-programme/doc80_en.htm (accessed February 2010).

Glover, D., Miller, D., Averis, D., and Door, V. 2005. The interactive whiteboard: A literature survey. *Technology, Pedagogy and Education* (14)2, 155–170.

Gonzalez, G.R., Organero, M.M., and Kloos, C.D. 2008a. Early infrastructure of an Internet of Things in spaces for learning. Advanced Learning Technologies, 2008. *ICALT '08. Eighth IEEE International Conference*. July 1–5, 2008, pp. 381–383.

Gonzalez, G.R., Organero, M.M., and Kloos, C.D. 2008b. Exploring touching learning environments. *IFIP International Federation for Information Processing. Learning to Live in the Knowledge Society*. Volume 281. Pages 93–96.

Google Maps. 2010. http://maps.google.com/maps?hl=en&tab=wl (accessed February 18, 2010).

ISTAG, Information Society Technologies Advisory Group. 2001. Scenarios for ambient intelligence in 2010. Advisory Group to the European Community's Information Society Technology Program. ftp://ftp.cordis. europa.eu/pub/ist/docs/istagscenarios2010.pdf (accessed February 18, 2010).

Lim, M.Y., Kriegel, M., Aylett, R. et al. 2009. Technology-enhanced role-play for intercultural learning contexts. *IFIP International Federation for Information Processing. Lecture Notes in Computer Science.* Entertainment Computing-ICEC 5709, pp. 73–84.

López-de-Ipiña, D., Vazquez, J.I., and Jamardo, I. 2007. Touch computing: Simplifying human to environment interaction through NFC technology. I Jornadas Científicas sobre RFID. Ciudad Real, November 21–23, 2007.

Luque, I. and Gómez-Nieto, M.A. 2009. University smart poster: Study of NFC technology applications for university ambient. *3rd Symposium of Ubiquitous Computing and Ambient Intelligence 2008.* Springer Berlin / Heidelberg. pp. 112–116.

Matas, G., Castro, P., Luque, I., and Gómez-Nieto, M.A. 2009. Encouraging learning and student motivation through NFC-based pervasive games. *Computers & Education.* Submitted for publication.

Matas, G., Luque, I., and Gómez-Nieto, M.A. 2009a. University of things: Applications of near field communication technology in university environments. *The Journal of E-working.* Vol. 3, 52–64.

Matas, G., Luque, I., and Gómez-Nieto, M.A. 2009b. How NFC can be used for the compliance of European Higher Education Area guidelines in European Universities. *Proceedings 1st International IEEE Workshop on Near Field Communication.* pp. 3–8. 2009.

Matas, G., Luque, I., and Gómez-Nieto, M.A. 2009c. Applications of near field communication technology in university environments. *Proceedings of the IASK International Conference.* E-Activity and Leading Technologies & InterTIC. pp. 127–134.

Moodle. 2009. Moodle: Open source community based tools for learning. http://moodle.org/(accessed January 2010).

Nava, S.W., Chavira, G., Hervás, R., and Bravo, J. 2009. Adaptabilidad de las tecnologías RFID y NFC a un contexto educativo: Una experiencia en trabajo cooperativo. *Revista Iberoamericana de Tecnologías del Aprendizaje.* 4(1), 17–24.

Newitz, A. 2006. The RFID hacking underground. WIRED. http://www. wired.com/wired/archive/14.05/rfid.html (accessed January 10, 2010).

NFC Forum Organization. 2010. Technical specifications. http://www.nfc-forum.org/home (accessed February 2010).

Päivi, J., Vili, T., Erkki, S., and Tapio, M. 2007. Improving mobile solution workflows and usability using near field communication technology. In *Ambient Intelligence*, ed. B. Schiele et al., 358–373. Springer-Verlag Berlin.

Ramirez, G., Muñoz, M., and Delgado, C. 2008. Exploring touching learning environments. *IFIP International Federation for Information Processing, Learning to Live in the Knowledge Society.* Volume 281. 93–96. Springer, Boston.

Roberts, C.M. 2006. Radio frequency identification (RFID). *Computers & Security,* 25: 18–26.

Rukzio, E. 2007. Physical mobile interactions: Mobile devices as pervasive mediators for interactions with the real world. Thesis, University of Munich.

Schmidt, A. 2006. A layered model for user context management with controlled aging and imperfection handling. MRC 2005, LNAI 3946, pp. 86–100. Springer-Verlag Berlin Heidelberg.

Smart Card Alliance. 2002. Contactless technology for secure physical access: Technology and standards choices. http://www.smartcardalliance.org/secure/reports/Contactless_Technology_Report.pdf (accessed February 18, 2010).

Smartouch Project. 2008. http://ttuki.vtt.fi/smarttouch/www/?info=intro (accessed February 2010).

Songdo project. 2010. http://www.songdo.com (accessed February 18, 2010).

VTT. 2009. Hot in the city. http://hic.vtt.fi/download.html. http://ttuki.vtt.fi/smarttouch/www/kuvat/December08_Newsletter.pdf (accessed January 10, 2010).

Weiser, M. 1991. The computer for the 21st century. *Scientific American,* 265(3), 66–75.

Weiser, M. 1993. Ubiquitous computing. *IEEE Computer,* 26(10), 71–72.

Zender, R., Dressler, E., Lucke, U., and Tavangarian, D. 2008. Meta-service organization for a pervasive university. *Proceedings of the 2008 Sixth Annual IEEE International Conference on Pervasive Computing and Communications.* 400–405. IEEE Computer Society.

Zender, R. and Tavangarian, D. 2009. Communications in computer and information science. Vol. 53. Intelligent interactive assistance and mobile multimedia computing. *Proceedings IMC 2009.* 73–84. Springer, Berlin.

7

NFC-Based Physical User Interfaces for Interactive Spaces

IVÁN SÁNCHEZ MILARA, MARTA CORTÉS ORDUÑA, JUKKA RIEKKI, AND MIKKO PYYKKÖNEN

Contents

Introduction

Near Field Communication (NFC) is a relatively new technology that evolved from RFID and proximity smart cards. So far, companies have focused on three different types of applications: Transit and ticketing, payment, and advertising. Transit and ticketing applications let users buy tickets, store them electronically, and use them later to access different premises and means of public transport. Payment applications, in turn, transform a mobile phone into a virtual wallet. Finally, advertising applications bring announcements and other promotional materials to mobile devices. All these application areas are interesting from the business point of view but cover only a small subset of the potential applications.

In addition to the applications listed earlier, RFID technology enables equipping any everyday object with a digital ID that distinguishes it in an unambiguous fashion from the rest of objects in the world. Moreover, NFC technology enables local communication between entities in our everyday environment. These features facilitate creating a network of interconnected objects known as the Internet of Things [32]. NFC focuses on user-initiated communication: because of the short reading distance, a user has to bring a reader and a tag near each other. As NFC technology becomes common, various electrical devices can be equipped with NFC readers. These readers can then be used to read and write RFID tags and even to communicate with other devices equipped with similar readers.

Our work focuses on mobile phones equipped with NFC readers, that is, NFC phones. Mobile phones are common in our lives, and an average user knows how to use such devices. Furthermore, nowadays even low-cost phones can connect to the Internet, making mobile phones the ideal devices to create links between objects in the physical world and communications networks. With RFID technology, such links can be created easily: a user needs just to touch with his or her NFC phone an RFID tag attached to an object. The digital ID and properties of that object can then be either read directly from the RFID tag or accessed from the network based on the data read from the tag.

NFC technology is a suitable technology for building physical user interfaces to control the services that an environment offers. When NFC technology is used, the interaction is realized by equipping objects with RFID transponders and touching these objects with NFC phones. Objects can be associated with events, with sets of events with service profile configurations, or with any other kind of digital information. An event can, for example, determine a command for an application. In the case of the multimedia player, touching a certain object can cause the player to advance to the next video, for example. A service profile configuration can determine, for example, the resources that a service should allocate before starting. This configuration depends on the environment context and the data setup in user profiles. The object might also have information about the services that can be controlled by interacting with it. Events and service configurations are sent to the target service when the corresponding RFID tag is touched with a mobile phone. The service allocates the necessary resources and executes the commands associated with those events.

We utilize NFC technology for building *interactive spaces*. Such spaces offer for users a rich set of services and easy interaction with these services. An interactive space contains resources such as displays, services using the resources, and physical objects for interacting with the services. In the rest of this chapter, we develop the concept of interactive space, propose an interaction model for interactive spaces, provide information on building interactive spaces using NFC technology and, finally, we provide a wide set of working prototypes that we have built using this technology.

Interactive Spaces

From Ubiquitous Computing to Interactive Spaces

During the last decade, the efforts of the research community in the areas of computer science, computer engineering, and HCI (human computer interaction) have been focused on achieving the new paradigm of hidden computation and calm technology that Mark Weiser envisioned 20 years ago [1]. This paradigm, which Weiser coined as *ubiquitous computing*, is defined by him as follows: "a physical world that is richly and invisibly interwoven with sensors, actuators, displays, and computational elements, embedded seamlessly in the everyday objects of our lives, and connected through a continuous network." *Pervasive computing*, *ambient intelligence*, and *everyware* [2] are other names that different research groups have given to ubiquitous computing. Although all these terms are almost synonyms, they have different shades of meaning. Ubiquitous computing has two main pillars: (1) every object in the environment should have computation capabilities and be connected to the network and (2) "calm technology," where computation and interaction "weaves itself into the fabric of our lives until it is indistinguishable from it" [1]. Hence, we can state the goal of ubiquitous computing is to hide all computation processes from the user and let the user interact freely with the environment instead of using classical input devices such as keyboards and mice. In other words, the goal is to enable computation and natural interaction everywhere.

The idea of bringing computation capabilities to the environment leads to the concept of smart spaces. In smart spaces, users interact with their environment just by performing actions (talk, move, sit) and using objects located in the environment. In a smart space, a user consumes services by interacting with the environment. This interaction does not disrupt her everyday life activities. A smart space is composed of three main elements: sensors collecting context and user information, actuators performing actions that modify the environment, and a computer system analyzing user behavior by processing sensor information and sending commands to the actuators. These commands modify the environment so that the user can consume the desired service.

The sensors in the environment are in charge of collecting context information about users and the space. Context has a very wide meaning, but according to Dey et al. [3], "context is any information that can be used to characterize the situation of an entity [...], a person, place, or object that is considered relevant between a user and an application." Based on context information (including user activity), application preferences, and user profile, the system decides the service or services that the user desires to activate and which are the resources he or she wants to use. For example, if the system detects that a person is sitting on a sofa in front of a TV and a match of his or her favorite sport is about to start, the computer system controlling this smart space guesses that the user wants to watch the match and decides to turn the TV on and tunes it in to the sports channel. Obviously, each space can provide only the services that it can support based on the place and the resources available. To offer the services, the smart space must be sensing continuously the environment using a wide range of sensors such as video cameras, movement sensors, microphones, pressure sensors, location sensors, and accelerometers. All those sensors are embedded in the environment or carried by the users. Furthermore, smart spaces need to process in real time the data produced by the sensors, abstract the information to create context information and, finally, activate a set of resources and services as a response to the user activity.

Schmidt [3] analyzes the main difficulties for user interaction in smart spaces. Interaction in smart spaces is either explicit, when a user interacts directly with a user interface, such as pressing a button; or implicit, when the system analyzes the context and user behavior to give an adequate response. Implicit human–computer interaction is the interaction of a human with the environment and with its artifacts, aiming to accomplish a goal. During implicit interaction, the system receives an implicit input. That input is not like the classical interaction with a computer but like the interaction that humans have naturally with other humans and with the environment. This type of interaction brings with it several problems, including: How to balance stability and dynamic usage concepts? How to keep the user in charge of the interaction and not end up with the user wondering about the actions taken by the system? A change in the context that the user was not aware of, or not correctly understanding the system behavior, might lead to changes in the environment totally unexpected by the user.

As natural interaction between humans and between human and the environment is complex, currently there are no computation models that represent this interaction completely. The challenges listed earlier have led us to take another path in developing ubiquitous computing systems: *interactive spaces*. While in smart spaces context recognition and implicit interaction prevail over other concepts, in interactive spaces explicit interaction is the keyword. The user always starts interaction with the system by performing an action in the environment. Context recognition is moved aside. Context is pulled by the system when the user starts the interaction, to create an adequate configuration that fits that context. This context pulling is another key aspect of interactive spaces. The system pulls the context after the user has deliberately initiated the interaction. The context is not pushed to the system, to analyze the behavior of a user who has not triggered interaction. Another difference is that while in smart spaces the computation is completely hidden in the environment, in interactive spaces the computation is integrated in the environment. The system advertises to the user the objects that he or she can interact with, how to perform this interaction, and what the expected results of that interaction are.

Interactive spaces emphasize user control; instead of studying solutions for deducing the users' goals automatically, we develop user interfaces that let the users communicate their goals to the system. Until the last decade, classical computer interaction was based on a computer with a keyboard and a mouse to give input to the system, and a display and a printer to provide output. Visual information was organized into windows, icons, and menus on the screen that the user can access using the keyboard and mouse. This style of interaction is known as WIMP (window, icon, menu, pointing device). This interaction requires the user to be in front of the computer, and hence this style of interaction is very far away from the idea of ubiquitous computing. In recent years, portable devices such as laptop computers, pads, and smart phones have changed this concept. These devices permit interaction with services everywhere, hence bringing computing a big step toward ubiquitous computing. However, the GUI is still based on the WIMP interaction style. Mouse and keyboard have been substituted by the touch screen in many cases, though.

Interactive spaces try to embed the computation in everyday life environments. Interaction with the system is not centralized in a single device, but every object in the environment can potentially give an input to the system. Tangible user interfaces [9] propose to represent digital content through physical or tangible objects that could be manipulated via physical interaction. This interaction style can be termed *tangible interaction* [27]. Objects are treated as tokens that can activate and control services in the environment when they are manipulated. That is, the objects provide access to digital information [28]. An object can have multiple digital representations. Each service selects the correct representation depending on the context, the characteristics of the service itself and on the type of interaction.

However, transition between what Valli calls natural interaction [5] and the classical human–computer interaction based on the WIMP style is not easy. Although natural interaction with physical objects in the environments is supposed to be the most intuitive way of interacting with a system, the years of using WIMP to interact with digital systems has shaped the expectations of the users. Many users probably think that the WIMP style is the only way of communicating with a digital system. User can be surprised when grasping an object in the environment produces an effect on a nearby display, although the metaphor is clear, and would be natural if the users did not harbor the WIMP prejudices. Until users overcome these prejudices, we should teach them how to interact with this new ubiquitous computing environment, creating a transition between the two interaction modes. For example, to indicate that the user can interact with an object to start that service, we can bring the WIMP concept to the ubiquitous environment and attach an icon to the object. The icon indicates the action that is performed when the user grasps, moves, or touches that object.

Researchers in the area of ubiquitous computing and smart spaces have developed systems that support interactive spaces. Abowd, in the Classroom 2000 project [29], builds a fully interactive space for learning environments, while Johanson et al. [30] show how to build ubiquitous computing rooms for collaborative work in office environments. Pering et al. has built Elope middleware [31], which provides a simple way of connecting computational devices to create ad hoc interactive spaces. However, those interactive spaces use only computation devices such as touch screens, PDAs, and computers to interact with the environment and not local resources.

We claim that NFC is an enabler technology for interactive spaces, since it permits creating a bridge between digital and physical worlds, and offers simple user interaction based on bringing one object close to another. Since NFC requires very close proximity (a few centimeters), we could consider NFC interaction as touch interaction. Later in this chapter, we explain how to create interactive spaces using NFC technology. However, first we need to define an interaction model for interactive spaces.

Interaction Model for Interactive Spaces

It is important to understand how users interact with interactive spaces. This is a prerequisite for designing interaction that maximizes user experience. Interaction with the services in an interactive space is radically different from the classical interaction with applications in a computer environment using the WIMP style. Although interactive spaces contain displays, interaction is not based on moving a pointer on a display and performing actions modifying the pointer's focus. Interactive space behavior has some differences to the classical behavior in traditional smart spaces. Here, the system does not sense the environment and explicitly infer context and perform actions. The user interfaces embedded in the space advertise the services available and the physical objects in the environment that a user can employ to interact with the services. The user performs deliberate actions with these physical objects to control the services, for example, shaking or moving the objects.

The classical interaction with desktop computers involves an interaction model that has been well defined for several decades. Similarly, in the last decade, much research has been put into the development of an interaction model for smart spaces [4,33–36]. These models are not totally suitable for interactive spaces. However, the similarities between smart and interactive spaces, make it possible to use some of the existing models as a basis when creating a model for interactive spaces such as the one proposed by Dahl [4]. He proposes five different design elements, namely: users, virtual zones, tokens, token containers, and computer devices. He also presents in his publication a set of icons to represent each design element, semantics for those elements, and examples of basic interaction.

Next, we summarize our interaction model for interactive spaces. The model is not fully complete, so we explain only the necessary concepts to grasp the main ideas on how to implement interactive spaces using NFC. Our interaction model contains five design elements, namely, *user, service, resource, mediator,* and *token.*

User is a person in an interactive space. This person can use the services offered in that space by interacting with tokens in the user environment (including the tokens carried by the person himself). Interaction with tokens is detected and communicated to the system through mediators. Interaction with other users is detected indirectly through tokens.

Service is a set of software functionalities and the policies that control its usage. Services can use the resources available in the environment. A user can interact with services through the tokens (resources can also offer traditional GUIs; see below).

Resource is any device located in the environment that services control. Usually, they are just output devices; that is, they provide feedback to the user. In that sense, they behave as actuators in classical smart spaces. Some resources can also be used as classical input devices. For example, some services could use a touch display to allow the user to enter data into the system.

Token is an active object in the environment. "Active" means that a user can give input to a service in the environment by handling the object. Each token is the embodiment of a digital object. Interacting with a token means interacting with the corresponding virtual representation, that is, the physical object's digital counterpart in the digital world. Files, commands, and configuration preferences can be represented as digital representations that are embodied by physical objects. This model dictates that the virtual representation is service specific, so generally it cannot be shared between services.

Mediator: In an interactive space, both token states and relations between tokens need to be recognized and communicated to the system. This is the task of mediators: they sense the state and communicate it to the system. Since objects do not usually have communication capabilities, we need some device that sends that information to the corresponding service, using any wired or wireless data bearer such as Bluetooth, Wi-Fi, Zigbee, etc. Mediators can also sense simple relations (such as "near"), either alone or together with other mediators. More complex relations can be recognized by integrating information from several sources. Some tokens have integrated mediators, but others require separate mediators to be attached onto them. For example, a mobile phone has an integrated mediator, but, for example, a wooden block requires a mediator such as an RFID tag to be attached to it. As another example, a Zigbee module detecting movements inside a soft toy is an integrated mediator. So, mediators have, in general, the task of sensing the change of state in the environment and transmitting this information to the system.

In addition to tokens, sensors carried by the users and installed in the environment can produce information about user activity. This information can be used by the system to infer the actions to be performed. Services can be advertised, for example, when a user enters a space. However, our focus is on user-initiated actions, on token handling performed by users.

In an interactive space, user interfaces are incorporated into the environment in such a way that a user interacts with different tokens in a natural fashion. The tokens that are part of the services' UI are integrated in the user environment. A user is not aware of the background services, but just uses the tokens letting the interaction flow naturally in the space. This is what Valli calls *natural interaction* [5]. However, we need a transition period in which we use some paradigms of the WIMP GUIs to provide correct affordances to the objects. Providing correct affordance is a prerequisite for a good user experience. A space must tell the user (1) which objects he or she can interact with, (2) the interaction mode of each object (grasping, squeezing, moving, touching, etc.), (3) what is the result of the interaction, that is, the information that is going to be sent to the service and how the service is going to respond (including the resources it is going to use to produce the response). Usually, the tokens are just used as an input device (output is usually provided by the resources). However, some tokens can also provide output information. Note that not only can objects in the environment be used as tokens, but also objects that the user carries, for example, a mobile phone or a watch. Another important point is that tokens can interact with each other. For example, placing a token near or into another can trigger sending commands to a service.

Sometimes, a resource acts as a token and represents the state of the resources as a physical object. In that case, the same physical object has two different roles. As a resource, that physical object has capabilities to perform tasks given by services, such as displaying information on the screen. As a token, it represents the state of the resource in the digital world. In that way, a user controls a service by handling the object as a token or changing its relation to other objects and a service gives tasks to the object as a resource.

As part of the interaction model, we propose three different stages in the interaction between the user and the services. These stages were initially sketched in [6]. The first stage is the *discovery* stage. During this stage, a user scans the environment and detects the different resources and tokens he or she can interact with. Since both tokens and resources can be at the same time input and output devices, we call both of them *artifacts*. Based on discovered artifacts, the user may desire to start a service among the ones he has found from the environment. In the *composition* stage, the user selects the artifacts he wants to use to control the service. Furthermore, during this stage, the user could also provide some input to the system. As a result, an application consisting of the selected service and user interface components (tokens and resources) is now ready for usage. Finally, during the *usage* stage, the user communicates with the application by interacting with tokens in the environment. The selected service uses resources to provide feedback or to perform the desired tasks. The tokens themselves can be used as output devices as well.

We can compare the different phases of the interaction model with the nowadays common process of downloading a video from the Internet and watching it using a desktop computer. First, a user searches a video on the Internet using some keywords, downloads the video into her computer, opens the player, selects the video to be shown, and uses the player interface to start the video. The Skål media player presented by Arnal [7] is the counterpart of this application in an interactive space. Skål allows starting and controlling a video in a nearby display by adding objects located in the environment to a bowl located close to the display. The command sent to the multimedia player depends on the object put into the bowl. We can analyze this application using our interaction model and compare it with playing a video from an Internet site. At the discovery stage, instead of doing a search by typing keywords, the user can visually scan the space to find real objects she would like to interact with. She might detect two displays in the room, two wooden bowls near the displays, and different toys distributed in the room. Since both bowls are physically connected to the displays, the user infers that every object she puts in a bowl starts a video in the corresponding display. In the composition stage, the user composes the service she wants to use: she selects the toy representing the multimedia content she would like to see and puts the toy into the bowl that is connected to the display she wants to use. In this case, each token (i.e., each toy) is the embodiment of a video. Moreover, the environment can contain tokens for controlling the multimedia player; wooden blocks for pausing the video, starting it, stopping it, etc. The toys and the wooden blocks contain RFID tags as mediators and the bowls contain RFID readers as mediators. The mediators detect relations "token in a bowl" that trigger sending commands to the multimedia player.

Our aim is to extend this model and define a language formalizing physical interaction in interactive spaces. This language will be used to describe the tokens available in interactive spaces and mapping between token handling (i.e., user interaction) and events sent to the system. This language creates a basis for creating applications from components distributed in the network and local environment, even dynamically based on users' context and selections. Hence, it will be one central enabler for interactive spaces. Although we have not completed this language, next we describe how to build interactive spaces using NFC technology.

Building Interactive Spaces with NFC

RFID allows building bridges between physical and virtual worlds in a simple way [8]. One RFID tag contains a variable amount of digital data. Passive RFID tags are small and flexible artifacts that can be embedded in almost any object. The data stored in a tag permits the system to generate a digital representation of the physical object, or related to the physical object. Furthermore, the data stored in a tag can contain configuration parameters or other kinds of object contextual information. By attaching RFID tags to objects, we can realize the tangible bits concept defined by Ishii et al. [9]. The tags are mediators enabling the status of the corresponding tokens to be detected and communicated to the system that can then generate and update the corresponding digital representation.

NFC brings three advantages to standard RFID technology: First, integration into mobile devices. An NFC reader chip is already embedded in some mobile phones. It is expected that in the near future, a great number of new phones will be equipped with NFC chips. Furthermore, there exist some software protocol stacks and APIs for NFC readers. Some programming languages used in mobile devices such as J2ME with its jsr-257, Android, and Python have their own implementations of the NFC protocol stack. Mobile phones support different kinds of communication technologies such as Bluetooth, Wi-Fi, 3G, and GPRS which can be used to transmit the data read from an RFID tag to any service in the Internet. Second, NFC permits full-duplex communication between two NFC devices. This opens interaction possibilities between mobile phones and objects in the environment. A mobile phone could, for example, transmit multimedia content to a display located in the environment when a user brings the phone near the display. Or, an object in the environment, with sensors incorporated, might transmit sensor readings to a mobile phone when the phone comes close to the object. NFC also supports an RFID tag simulation mode, which permits using a phone as an RFID tag. It is the same to attach RFID tags to the phone's owner. Third, the NFC Forum is standardizing NFC technology. Standards guarantee interoperability between devices from different vendors. NFC standardization effort does not cover only the protocols but also formats for storing data in tags.

NFC technology is a great tool to build the interactive spaces described in the previous section. Nearly any object in the environment can be transformed into a token by attaching an RFID tag onto the object. The tag contains the data that the system uses to create a digital representation of that object. The same object can contain several RFID tags, for example, at different parts of the object, or to transmit different states of the same object to the system. For example, a puppet could have one RFID tag in each hand and one RFID tag on its forehead. The meaning of touching a hand with the mobile phone would be different from touching the forehead, so the information transmitted to the system would be different. Furthermore, objects with computation capabilities could be equipped with NFC reader chips allowing direct communication with mobile phones near the object. These readers could be used to realize different interaction modes. For example, the token could send different information if it is squeezed or if it is shaken.

Some resources can also have RFID tags. These tags are usually used in the composition stage of our interaction model to (1) give information to the user about the resource and (2) make known to the system that the user desires to interact with this resource.

As mobile phones are capable of transferring the content read from an RFID tag or an NFC chip to the server running the service, they can be used as mediators. Mobile phones are good choices, as they are widely used around the world, and hence well known by users. Moreover, as a mobile phone is a personal device, it can store private data, such as personal preferences and user ID. This personal information can be transmitted when required to the interactive space system after user acceptance. Furthermore, a mobile phone can be used as a tool to configure the interactive space when the objects in the surroundings space will not suffice. Mobile phones can also provide to the user feedback (visual, auditive, and haptic), helping him or her to understand whether the interaction has been performed correctly. Finally, using the phone as a mediator requires the user to perform explicit actions (to touch tags). The system controlling the local environment does not sense implicit input (as generally happens in smart spaces) but waits an explicit input from the user. So, the system only provides an output when the user touches an icon, that is, when the user requests output. Such an operating mode makes the system more efficient, less prone to errors (provides output only for explicit input), and improves the feeling of the user being in control (something happens only when he or she desires).

The services are distributed into the Internet. The services receive inputs from the mediators and directly from the tokens (which have integrated mediators). The input usually contains a command and some parameters. Those commands and parameters are usually stored in the RFID tag. However, a mediator (mobile phone) can filter or augment the data, for example, add parameters related to the user to personalize the service. Additional data can also be fetched from the network.

We can analyze the NFC-based user interaction in an interactive space using our interaction model presented earlier. Sanchez et al. [9] describe in further detail the requirements to create this kind of spaces using NFC technology. The interaction process starts when a user enters an interactive space. During the discovery stage, the user scans the environment, searching for the services, resources, and tokens that can be used. In an NFC-only interactive space (i.e., no other interaction methods), an icon and an RFID tag are attached to every token and resource a user can use to communicate with the services. An RFID tag, placed under the corresponding icon, stores the data to send to the system when the mobile phone touches the icon. Each icon contains a pictogram communicating to users the affordances of the corresponding object, that is, the properties that determine how the object can possibly be used to control services. We adapt the perceived affordance concept defined by Norman [37]. As we mention in the previous section, the same object can have different usages in the interaction. In this case, the active object contains several icons, each one identifying the function of a specific part. The next section discusses locating icons on objects and the appearance of icons in more detail. In addition, for example, posters attached to the walls of the space can provide additional information on using the services and the physical user interface, especially when the users are practicing this new interaction concept.

During the composition stage, the user selects the resources and tokens he wishes to use in his application. During the usage stage, some other resources and tokens could be used, but the core of the application is created during this composition stage. The user expects the output of the system to be produced by the resources, while the tokens might be the representation of configuration parameters, multimedia content, other type of files, and the commands. Resources can be selected either automatically by the system, selected from a list presented in the mobile phone's screen, or by touching icons associated with resources [10]. To present a list, the mobile phone needs a resource discovery application to find the available resources in the space. This list could be ordered giving the best combination of resources based on user preferences, service to run, context information, and the current state of the resources and system. Davidyuk et al. [10] present algorithms and example applications to perform this task. A user can interact also with the other users present in the same interactive spaces. A user can touch other users' mobile phones or RFID tags that identify the other users to interact with.

Finally, during the usage stage, the user controls the application that he or she has just created during the composition stage. In an NFC-only interactive space, the user controls the application by touching icons placed in the environment. Each icon is associated with one command or a set of commands. Some applications can be controlled just using the mobile phone GUI (keypad and display).

Although we focus on NFC-only interaction, mobile phones introduce the possibility of multimodal interaction with the application. For example, an integrated accelerometer enables gestures, and the phone's microphone speech input. Furthermore, an icon can be placed on top of an RFID reader. In this case, the icon is not bound to one service, but the system can change dynamically the data that is read into a terminal when the icon is touched; the icon needs to advertise this to the users. When an icon is placed on top of an RFID reader, the mobile terminal does not have to be equipped with an RFID reader, but an RFID tag suffices. When the user brings the terminal near the reader, data identifying the terminal is read from the tag and delivered to a server.

Advertising Icons

An icon advertises a point in the environment that can be touched with a mobile phone and the input sent to the service when this icon is touched. An icon forms, together with the RFID tag placed behind it, a two-sided interface between the physical and digital worlds. The icon advertises an action to a user and the tag contains data sent to the system when the tag is touched. Although the potential of this kind of NFC-based interfaces between the physical and digital worlds is considerable, research on advertising RFID icons is surprisingly rare. Some iconic representations have been proposed by Arnall [11], Tungare et al. [12], and Valkynen et al. [13].

In this section, we describe the main challenges on icons advertising RFID tags, how we cope with these challenges, and how icon design has evolved over the years. An icon (1) must communicate to the user that it is part of the interactive space's user interface, and (2) must be designed in such a way that the object's affordance perceived by the user is the same as the affordance sought by the designer. When these requirements are fulfilled, the user understands that he can interact with the space by touching the corresponding object. If users do not recognize all icons that belong to the interactive space, some services will not work as expected. On the other hand, if a user recognizes an icon but interprets it incorrectly, the output sent by the system will not be what the user expected, which would detract from the user experience. Moreover, if the user touches an icon that is not part of the interactive space, there would be no output from the system and the user might think that the system is broken.

To tackle this challenge, we have divided icons into two parts: the outer part is a general icon that communicates to the user that a point can be touched. The inner part, in its turn, contains a pictogram that is a metaphor of the input that the system receives when the tag is touched. The inner part can contain text as well. For example, we might place an icon with a pictogram "say out loud" on a toy. A learning language application would then say the name of that toy out loud when a user touches the tag. On many occasions, the icon represents a command to send to the system but, in some cases, it informs the user about an object that can be touched. In such cases, the outer part alone or a specific inner part communicating just a point to touch suffices.

Sometimes an icon cannot be placed directly on the object we want to interact with, since this object is not accessible by the user. In that case we can place the icon in a representation of this object, such as a photo or a poster. For example, when using a photo album service, a user could filter photos by the persons that appear in them. In this case, icons could be attached to photographs of the persons. Generally, this approach requires a mapping between the real object we want to interact with (e.g., a person) and the representation of that object (e.g., a photo).

Figure 7.1 offers an example icon of the first iteration (left) and an example icon of the second iteration (right).

We are now in our third iteration of icon design. Our icon design has so far focused on starting services by touching RFID tags. Furthermore, we have linked objects directly to one service, so the same object has not been used for several services at the same time.

Figure 7.1 Examples of icons.

Our first icon design for interactive spaces was presented by Riekki et al. [14]. This design contains two types of tags, a general tag and special tags. Special tags identify the specific information that a user gets from an object when a tag is touched. The general tag icon has a rectangular shape representing an RFID antenna. Special tags have the same shape, but in addition are augmented with an action symbol at the bottom-right corner. The same color is used to represent the general icon and the action symbols.

In the second iteration, presented by Riekki et al. [15], we improved the general aesthetics of the icons. We changed the shape of the icons from rectangular to circular. The external border is framed with a blue and black band that is the same for all tags in the environment, and communicates to the user that this icon can be touched. To facilitate the work of the developer, the ID of the tag is written in small fonts in the frame. The internal part of the icon represents an action. Sometimes, for clarification we include some text in the internal part. Furthermore, as this technology is new for the majority of users, we have embedded some of the icons in posters, which instruct the users how to use the application. For some services, we have used both methods; first, we explain to the user how to use the application in a poster, and when the user is familiar with the application he can use the icons located in the environment.

After the second iteration, we realized that some users did not understand how to start the application, and many of them were surprised when we explained to them that they did not have to start any application manually using the phone's GUI, but just touching the icon would bring the application to the mobile phone and to the environment. The results of our usability experiments were in line with the results of Broll et al. [16]. They performed a usability test to improve the learnability and guidance in NFC interaction. Among the main discoveries of the usability tests, it is worth pointing out the following ones: (1) having a start tag in a poster that explains how to start the interaction facilitates interaction, (2) physical objects should have visual cues indicating that the objects can be interacted with, and (3) visual feedback and instructions should be provided on the physical object rather than on the mobile phone's screen.

Our goal is to build a system that can be used at any time by any user, without any instructor nearby. Within ubiquitous computing, the interaction is usually represented using alternate modeling techniques such as storyboards and sketching instead of formal UML representation [17,18]. We use the same idea in our third iteration. In this design, comic strips placed in posters in the environment (1) explain to the user how to interact with the environment using NFC and (2) contain the icons to command services. The icons are placed next to the corresponding actions in the comic strip. Those comics strips are placed close to the object or resource that the actions are related to. Due to the size of the comic strip, it is not easy to embed it in the real object. However, we think this is a great tool to instruct users until they understand how this new technology works. Figure 7.2 shows at the bottom one comic strip that instructs the user how to use an application. The upper part contains a storyboard that gives access to five different services. The services are explained in the section dedicated to new applications later in the chapter.

The icon design is similar to the second iteration, but simplified. The circular frame is thinner, and to avoid confusing the user we do not include the RFID tag numbers. Furthermore, we found that sometimes the users did not know where the output of the interaction was going to be represented. That is why we added at the bottom-right corner small icons indicating the devices providing the output (displays, speaker, mobile phone, etc.).

Figure 7.2   A comic strip instructing users how to use the available services and containing icons for controlling the services.

Interaction and Feedback

The user must always get feedback after touching an RFID tag. Since we use a mobile phone as a mediator, the feedback must be provided using the output capabilities of the phone. We propose using a combination of haptic and audio feedback to indicate whether a tag was read successfully or some error occurred. We achieve two goals with this feedback: (1) users know if the input was transmitted successfully to the system, and the interaction does not have to be repeated and (2) users receive confirmation that the area touched with the mobile phone is an active area, so output can be expected. If there is no feedback at all after touching an icon, the user learns that the touched area is not active.

In our usability tests, haptic feedback was valued by users, mainly because of its nonintrusive nature. Sound feedback was not studied in detail in these usability tests. We agree with Broll et al. [16] in that important output requiring a user's focus of attention should not be provided by the mobile phone. This is related mainly to visual output, as well-designed haptic and audio feedback can be observed at the periphery of attention, while focusing on something else. The reason for this recommendation is that a user, when utilizing tangible interfaces, expects the output to be provided by external tokens and resources, not by the mobile phone. The mobile phone is just a tool, a kind "magic wand" used to interact with the objects in the environment. Furthermore, the multimedia capabilities of the phone are usually worse than those of external resources.

We propose the use of the mobile phone screen only in three situations: First, error messages can be shown on a mobile phone's screen when no suitable external resources are available or error messages cannot be transmitted to the available resources. For example, if a user would like to start a service at the composition stage but some configuration needs to be done before the service can be used, a message can be shown on the phone's screen. Second, a mobile phone's screen can be used when the GUI of the mobile phone is directly used to send input to services. For example, a multimedia player's GUI (commands to play, pause, stop, move to next, etc.) could be shown on a mobile phone's screen. In this case, a user can select a command using a mobile phone keypad or a touch screen. Third, a phone's screen can be used when there are no external resources available to show the expected content. However, visual feedback on a mobile phone's screen should always be accompanied with haptic or audio feedback drawing the user's attention to the screen. The information showed in the display should be minimal. It is preferable to use graphical icons and animations rather than text. If text is needed, the message should be as simple as possible. Finally, the mobile phone keypad should be avoided. Interaction should be done with the environment.

Data on Tags

Where should data be stored? Should all data be stored in RFID tags, should the tag contain only an ID and access the rest of data using the network, or is some compromise between these extremes the best solution? These questions have been studied by Diekmann et al. [19]. As nowadays the storage capabilities of the commonly used tags are quite constrained (several kilobytes at maximum), we cannot think of storing all data in tags. Low transfer rates also hinder reading large amounts of data from RFID tags with NFC phones, as the user has to keep the phone in place during the read operation. On the other hand, reading data from the network increases the complexity of the system and the latency. Hence, we need a balance between data stored in tags and data stored in the network.

A tag should contain all data needed by the mobile phone to generate the corresponding input to the system. The phone can use also some additional data, stored in the phone when generating the input. The data stored in the tag should be sufficient, for example, to detect that the touched tag is not available for the current service. Some services do not require any information to be sent to the network; for example, business cards picked from conference badges can be stored directly in a mobile phone's address book. In this case, all data is stored in tags. When data must be sent to the system and a network connection is not available, the phone needs to inform the user clearly about the error. When possible, the phone should store this data, and send it to the system when the network is available. In some cases, data utilized by a service can be changed by the users of the interactive space; in that case it is advisable to store as much data as possible in tags. Of course, network storage has to be used for large amounts of data. If the content of the tag should change dynamically and very often and only the application administrator can change that content, then is advisable to store the data in the network. However, if the data is static, then it is again advisable to store data in the tag.

The function of a tag depends on the services in which the token is to be used. A tag (and its icon) is generally a metaphor of one action that can be performed with the corresponding token. Data is stored in tags following the standards provided by the NFC Forum. A tag stores the data in a NDEF message [41]. Each chunk of information is stored in a NDEF record [42]. The system recognizes the kind of data stored in the record type. We propose the structure shown in Figure 7.3 for NDEF messages in an interactive space.

The NDEF message should have at least three different NDEF records. The first one identifies the application that should be opened in the mobile phone. That is, the first record links a tag to a mobile phone client. This NDEF record does not contain a payload, since

NDEF record Type: Service dependent	NDEF record Type: NFC URI	NDEF record Type: Action type	...	NDEF record Type: Action type

Figure 7.3 NDEF structure for messages in interactive spaces.

the mobile phone just needs the type of record indicating the application. The push registry of the mobile phone (if using MIDlet-based clients) should specify the application that is to be launched when this record type is read. The second NDEF record has the NFC URI type. This URI identifies an Internet address to download the mobile client Over the Air, if the application determined by the tag is not installed in the mobile phone. Finally, the NDEF message contains an array of NDEF record types that encode the different bits of information associated with the token. Based on our prototypes, we have identified seven different kinds of record types to build the token information. There is an eighth record type used to build some of the other seven named Parameter. It is an extension of the NFC Forum Text Well Known Type. The payload contains text with the following format: *'parameterName'='parameter Value'*. Based on the prototypes we have built, we have identified the following NFC Forum external records types:

StartService: This type is used during the composition stage, and it provides information related to the name of a service and its configuration parameters. This is one way of providing context information to a service. The same service can be started from different places in the same interactive space. Each place can have its own configuration parameters stored in a tag. This type is formed by an NDEF message described in the Figure 7.4.

The first record of type NFC Text defines the name of the service, in US-ASCII. The rest of the records are (parameter name, parameter value) pairs defining the initial configuration of the service.

Command: This type stores a command or a set of commands that are sent to a service when a tag is touched. Usually, the same command tag can be valid for different services. In addition to a

NDEF record Type: NFC Text	NDEF record Type: Parameter	...	NDEF record Type: Parameter

Figure 7.4   NDEF message structure for StartService.

command name, parameters can be stored in a tag. The format of the record is the same as the one defined by the StartService in Figure 7.4. The only difference is that in this case the first NDEF record identifies the name of the command. One RFID tag can contain several records of type Command.

State: This type contains state information of an entity. For example, for a time-tracking application, a user needs to touch the "Enter" tag when arriving at work. This causes the state of the user to change from "out of work" to "working." The type is structured as an NDEF message formed by two records: the first one is of type NFC URI determining the ID of the object, while the second one encodes the new state in an NFC Text type.

FileStorage: This type is used as a file repository. Users can "pick" and "drop" files to the tag using this type. The file is not physically stored in the tag because of the limited size of data storage. Instead, a tag contains a reference to the file. Each file in a tag is stored in a separate record. Each record contains a reference to the file as well as extra metadata such as file size, file type, and textual description. Ideally, each piece of metadata should be stored in a different NDEF record. However, due to the small amount of data that an RFID tag can contain nowadays, both the file ID and its metadata is stored in only one record of type NFC Text. The format of the payload of this record is as follows: 'file_uri'|'file_mime_type'|'file_size'|'storage_date'|'file_owner'|'file_description.'

PickAndDrop: A tag containing this type allows picking (dropping) content from (to) the tag. The main difference from the previous type is that the previous one represents permanent storage and this one, temporary storage. The action is valid only for the current session. The tag contains a reference to a service, to a resource, or to a combination of service and resource showing or storing the content. In the case of picking, the mobile phone picks content that is currently in use by the resource or service, while in the case of dropping, the content selected by the user is automatically moved to the target resource or service. The format of this type is an NDEF message that contains an array of records of type NFC URI. Each one identifies a resource or a service showing or storing the content.

Entity: This type represents the object to which the RFID tag is attached. The data stored in the tag is a URI, which identifies the object, a set of records of type NFC Text, which are used to give the name of the entity in different languages, and the metadata required by the particular application. This type can be used also to communicate to the user the characteristics of the device on which the tag is attached. This type can be used by various services that handle information about objects, for example, logistic services. We describe below an application that says out loud the name of an object when a tag is touched. The format of the NDEF message is shown in Figure 7.5.

This type can also be used to create communication pipes between different devices. Each device has a tag containing a record of this type. A user selects the ends of the pipes by touching those tags. The system creates a pipe between the selected devices, using a technology that both devices support. When a communication pipe is created to the mobile phone, a tag in the other device is sufficient. Linking functionality can also be integrated with other actions; for example, touching a "Print" tag might trigger a Bluetooth link to be created between the mobile phone and a printer.

Location: This type stores the coordinates of a place as well as a name or ID.

NDEF record Type: NFC URI	NDEF record Type: NFC Text	...	NDEF record Type: NFC Text	NDEF record Type: Parameter	...	NDEF record Type: Parameter

Figure 7.5   NDEF message structure for Entity.

Apart from the NFC record type format described in this section, for some applications we have used directly the Well Known Record Types defined by the NFC Forum, as well as some MIME types encoded in a tag. For example, we have used the URL record type to store in a tag the URL to download Over the Air some applications to the mobile phone. We have also used some Nokia-specific tag formats such as the one used to store an SMS or a phone number in an RFID tag.

System Implementation

We have built a system called REACHeS for building physical user interfaces to interactive spaces. In this section, we sketch its most important characteristics. For more detailed information, we invite the reader to check some of our previous publications [15,20,21]. The REACHeS platform is not yet complete. The current version does not support all the functionalities needed for an interactive space nor all tag types described in the previous section. REACHeS is a server-based platform that allows communication between Internet services, environment resources, and mobile clients acting as remote controls for both services and resources.

REACHeS platform has the following features:

- Resource and service registration.
- Resource allocation for services. Once a resource is allocated for a service, the service can control the resource by sending commands to it. Resource allocation can be done automatically by the system or by using resources selected by the user.
- Command redirection. Mobile clients send commands to the REACHeS server, which processes the command and forwards them to the corresponding service.
- Control of mobile phone's GUI. A service can modify a mobile phone's GUI dynamically. This feature enables creating on the phone display a remote control interface for the service.
- Client session control.
- Error control.
- Mobile client application supply Over the Air.

The system overview is shown in Figure 7.6. A user requests a particular service by selecting the corresponding icon, and touching it with

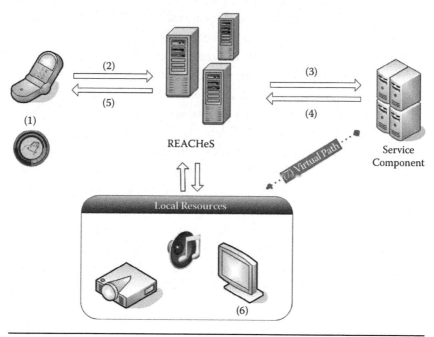

Figure 7.6   The REACHeS system.

a mobile phone. The RFID reader installed in the phone reads data from the RFID tag placed under the icon (number 1 in Figure 7.6). The mobile client processes the data read from the tag, filters it, and sends an HTTP request containing the data to the REACHeS system (2). REACHeS filters the request and forwards it to the server that is responsible for the service in question (3). The service replies with a message that determines the user interface to be created to the mobile phone (4). REACHeS adapts and passes this information on to the phone (5). As a result, the mobile phone GUI is changed to show a system message or a GUI to control the service. Until the service is closed, REACHeS transmits messages between the mobile phone and the service (as indicated by 2–5). The service can also request the control of local resources (e.g., a wall display) from REACHeS (6). In such a case, REACHeS forwards asynchronous commands from the service to the local resources. From the service's point of view, the events flow directly to the resource (7). In a typical message sequence, a command given by the user is first delivered through REACHeS to the service. The service then responds by sending a response to the mobile phone and a command to a local resource. The communication process proceeds like this until the service is stopped.

The sequence diagram in Figure 7.7 clarifies the behavior of REACHeS. Before the system can be used, services and resources must first be registered into REACHeS. When registering a service, the information required to access the service (host name, address, path, ...) is delivered to REACHeS. Additional parameters such as display location can be provided and used later in resource allocation.

Communication between the four entities in the figure is performed using HTTP GET requests. Messages sent between the mobile client and REACHeS and between REACHeS and the server contain a service parameter that indicates the target service, a command parameter that indicates the command to be sent to the service, and a set of (parameter name, value) pairs that are processed either in REACHeS or in the target service. When REACHeS receives a message from

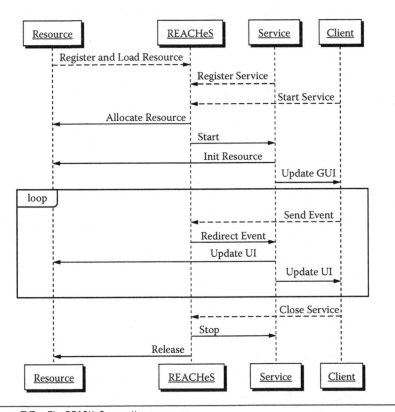

Figure 7.7   The REACHeS operation sequence.

a client, it filters the parameters, adds new parameters if necessary, and redirects the resulting message to the target service. The service replies with an HTTP response. The body of the response includes information to generate or modify the UI in the mobile phone. This response is processed again by REACHeS, adapted to the type of client device, and then sent to the mobile client.

When a resource is loaded, REACHeS opens a data stream to it. The services can send asynchronous HTTP POST requests to REACHeS to control the resource. The request body contains the commands to be sent to the resource. REACHeS keeps in its database a list of resources for each service. REACHeS reads from a request the commands, and forwards them to the resources using the data stream between the resource and the system. The resource processes the commands, and modifies its state accordingly.

The server is implemented using Java Servlet Technology and runs in Tomcat 6.0. A chain of servlets processes the request from the mobile client and forwards them to the corresponding service. Each type of resource (display and speaker at the moment) has its own servlet controlling all resources of that type. Other servlets are in charge of specific tasks such as allocating resources, updating the database, and administrating the system. A set of JSPs offers a GUI to administrate the whole system. Display and speaker resources' software runs on a web browser. Content is rendered using web pages and Flash technology. Communication protocols and real-time control is performed using Javascript.

Mobile clients are running on Nokia 6131 NFC devices. Those devices offer an implementation of the JSR-257 API to access NFC functionality, so all mobile clients use J2ME MIDlet technology. The server side offers a MIDlet repository. Mobile devices can download and install the MIDlets Over the Air. There is no single MIDlet to control all the services in the environment; that is, each service has its own mobile client. Some MIDlets can be suitable for controlling more than one service. If a mobile phone does not contain the required MIDlet, a tag in the environment should contain a link to download and install that application over the air. The MIDlet that the phone opens depends on the type of the first NDEF record stored in the tag. Several services can have the same type of share, that is, a MIDlet. We have implemented a general library that allows a MIDlet to read data from a tag and send it to REACHeS. Each MIDlet has to realize an application-specific state machine as well as the GUI. The content of the NFC tags are explained in the previous section.

The current system is based on a centralized server approach. The system is flexible enough for interactive spaces as services can be running on the Internet (and not in the REACHeS central server), and a great number of resources can be loaded in the system at the same time. However, a system to support interactive spaces should be more distributed. Currently, we are researching other alternatives for the infrastructure. Our first more-distributed implementation is presented by Sanchez et al. [21]. This botnet-inspired system uses the existing IRC infrastructure as a core system. Each entity in the system (mobile client, resources, and services) implements an IRC bot providing a communication link to the other entities. Services and resources open IRC channels to communicate with other IRC bots. Other IRC channels are used for administrative purposes such as resource and service discovery. Using the same botnet idea, we can create botnets that communicate with each other using the XMPP protocol [38]. Entities can then communicate with each other using the existing XMPP infrastructure. We are studying also the smart space middleware as the possible interactive space communication platforms. Smart-M3 [39] is one such candidate.

We are planning the following improvements to our REACHeS system: changing the lower layers (communication layers) to some of the alternatives listed earlier, offering a simple API based on RESTful architecture [40], using a distributed database, using a NoSQL approach more adequate for interactive space than classical SQL, and using the publish/subscribe paradigm for the communication between services and resources, which should be more versatile and more efficient in terms of network traffic and load.

Applications

The following section lists the prototypes we have implemented so far. The interaction model described above was built based on the experience gained during developing and testing these applications. For each application we describe the application and the tokens, resources, and mediators used.

Touch & Control

Touch & Control is the first NFC-based application built for interactive spaces that we are aware of. Sanchez et al. [22] present this application for starting and controlling multimedia applications on wall displays. Mobile phones and RFID tags placed in the environment form a remote control for the application. Each RFID tag is associated with a command and is placed under a control icon advertising the command. When a user touches a control icon, the corresponding command is sent to the service in question. The service processes the command event and performs the requested action, for example, updates the display content.

In our first prototypes, we used RFID tags just to start services. In these prototypes, each tag is associated with a different service. Touching a tag with an NFC phone triggers a start event that is sent to the central server (REACHeS platform). REACHeS recognizes the target service, allocates necessary resources (such as displays), and loads to a GUI the user's mobile phone for controlling the service. Using the mobile phone keypad, the user can, for example, send to the service a command to increase the volume of the video that is currently being played on a display. After some usability tests, we realized that users did not feel very comfortable using the mobile phone GUI to control services and displays. We used the Nokia 6131 NFC mobile phone, which has a display resolution of 240 × 320 pixels and a size of 2.2 inches. Users criticized the difficulty of controlling the service using the mobile phone due to the small sizes of the screen and the keypad. The users had to learn to use the mobile phone GUI. Moreover, two different focuses of attention (phone display and wall display) increased the cognitive load.

The solution to this problem was to embed the user interface in the environment. Instead of using the tags only to start services, we built a complete user interface from RFID tags. This touchable control board contains a control icon for each command of the service.

RFID tags are placed under those icons. When a user touches an icon the corresponding command is sent to the service via REACHeS. For example, if a wall display is playing a video a Pause icon on the panel can be touched to pause the video. That is, users do not have to use the mobile phone keypad to send commands; they just need to touch the appropriate icon with the mobile phone. From the user point of view, a tag is like a mechanical button of a classical remote control. When the button is touched, a command is performed and the user can see the effect.

This approach has several advantages in addition to releasing the user from observing the mobile phone's screen. First, as we are not restricted by the size of the mobile phone screen, we can have big icons. This is especially useful for visually impaired people. Second, as icons can be located freely, we can embed the remote control application completely in the environment. The UI can be placed wherever it is accessible by the users; on walls, posters, books. We can even associate a command with different objects in the environment. For example, we can create a user interface for children in which they can control their favorite animated films by touching different toys with their mobile phone. Third, the requirements for infrastructure are minimal; just the icons advertising the tags, the tags, and a display with an Internet connection is needed. User interfaces can be personalized easily: changing the GUI means only changing the icons that are shown to the user. If the icons are shown on a single poster, printing a poster with new icons is enough. So it is possible to change the application GUI without any programming, compiling, and downloading new software versions.

As an example application, we built a multimedia player for watching videos on a wall display in 2007. A start tag stores the list of videos to show in the session. In addition to the start icon, the control panel has icons for five commands: play, pause, stop, next video, and previous video. Figure 7.8 shows four different control panels for the multimedia application. Figure 7.8a shows a control panel embedded in a poster. A transparent version of the same control panel in a shop window is shown in Figure 7.8b. Figure 7.8c shows the control panel embedded in a CD case, while Figure 7.8d shows the same control panel folded into a 3D object, a cube.

(a)

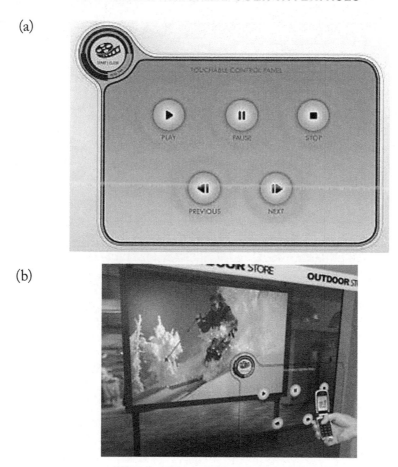

(b)

Figure 7.8   Four different control panels for the multimedia application.

Since the user interface is distributed in the environment, the GUI shown in the mobile phone screen can be minimal. It just shows instructions to touch icons. When a tag is touched, the command sent to the system is shown on the phone GUI. Also, error messages such as network problems and incorrect tags are shown on the phone GUI. Finally, to provide better feedback to the user during the interaction, haptic feedback is provided when an RFID tag has been read successfully.

Another example service using Touch & Control has been reported by Turunen et al. [23], where the authors built a media center that can be controlled using different input methods, among them NFC technology. The media center controls the EPG (Electronic Programming

(c)

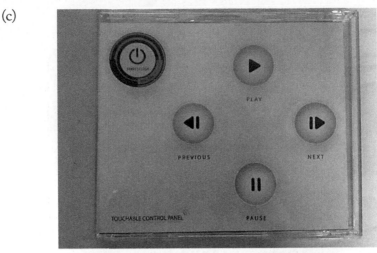

(d)

Figure 7.8   (Continued)

Guide) of a digital TV. The EPG user interface is shown on the TV screen. It consists of a grid, where columns represent television channels and rows represent time slots. Cells are individual television programs. Each cell shows an icon representing a type of program (sports, series, documentary, news, etc.). The system also can schedule the recording of a specific TV program and play recorded programs.

We implemented an NFC-based user interface for controlling the EPG. The Control Board built to control this application is shown in Figure 7.9.

The main difference from the previous application is that in addition to instant commands (shown in green at the table: zoom in, go

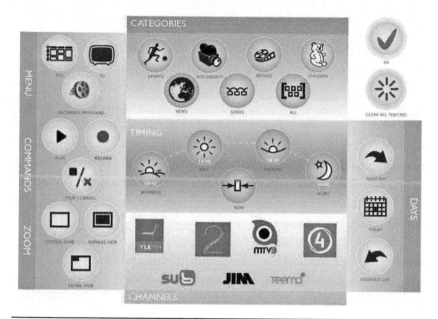

Figure 7.9   The control board for controlling the EPG.

to the next day, etc.), some commands can be stored in the mobile phone, in a stack to form composed commands. For example, a user could choose to highlight all sport programs (category, in blue) scheduled for the morning (time, in pink) from channel 1 and 2 (channel, in yellow), by touching the corresponding icons in the board. The composed command is sent when the user selects the Send command from the mobile phone's UI or touches the OK icon (top right). The mobile phone display shows in a stack all the icons touched until the composed command is sent. The main challenges we faced in building the Touch & Control application were related to usability. However, the usability test of this prototype showed that users accept a control board with a large number of icons, if the visual representation is clear enough.

Touch & Control uses two different NDEF record types described in the "Data on Tags" section: StartService and Command. The tokens are the different control boards shown in Figures 7.8 and 7.9. Some of them are embedded in physical objects such as the one embedded in a CD case or in the cube, while other tokens are the control boards, as in the case of the EPG controller.

Touch & Share

Touch & Share [24] is an application for sharing content in our everyday environment. The content is stored in local file containers and personal file containers. Visual icons placed in the user environment act as local file containers. Data is physically stored in RFID tags placed behind the icons. A personal container is a database stored in a user's NFC phone. A copy of a personal container can be maintained in an external server, so a user can access his or her personal container from many different devices. Content is shared between users and the environment by transferring files from local containers to personal containers and vice versa.

Users interact with the local containers using the Pick & Drop paradigm proposed by Hosio et al. [25]. Pick and Drop actions are initiated by touching a local container with an NFC phone. The first action, Pick, transfers files from a local container to a personal container, whereas the second action, Drop, transfers files to the opposite direction. The RFID tags do not contain actual files, but references to locate the files from the Internet, and some metadata. However, a user sees the files being transferred directly from the local environment to a mobile phone and vice versa.

The icons are placed on, or near, the physical objects the files are related to, or on a representation of that object. For example, a name tag could act as a representation of a person. The relationship between the files stored in the local container and the object containing the icon depends on the application and on the context. For example, if a local container (i.e., an icon) is placed at the entrance of a meeting room, the container could store the presentations of a meeting arranged at that room. The participants of the meeting could then pick the presentation files to their personal containers. As another example, a container icon placed close to an artifact in a museum could store some multimedia content related to that artifact.

Each tag of Touch & Share contains an array of NDEF records of type FileStorage. Each record represents a file. The tokens are the objects where the tag is located or the representation of that object. In this case, sometimes the tokens are not directly manipulatable by the user. For example, a museum artifact would not be accessible. However, the icon that represents this artifact is placed close to it.

TiPo TiPo [26,24] is a prototype based on Touch & Share. It was installed in the Zoological Museum of the University of Oulu during 2008, and tested by more than 300 pupils of different schools in Oulu. The museum has a collection of stuffed animals. We enhanced 22 of them by attaching a local container icon near each animal. The containers store images, sounds, and web pages related to the target animal. Visitors pick this content with NFC phones during their visit to the museum. They can later play (listen and watch) the collected material using the mobile phones or desktop computers. A web application on a desktop computer accesses a user's personal container in the central server. We constrained the functionality so that only administrators were able to drop files to the local containers, and museum visitors could only pick files from the local containers.

CADEAU

CADEAU is a prototype developed during 2008 and 2009, helping a user to compose an application from services and resources [10]. The goal of this prototype is to study the composition stage in interactive spaces. To be more specific, the goal is to study the combination of autonomic mechanisms and user interaction at the composition stage. Three different alternatives were implemented: In *autonomous composition*, the system performs the composition without user intervention. When a user starts an application, the system selects the best combination of resources and services available for the application. The selection is performed by genetic algorithms, which take into account the functional and nonfunctional properties of the resources. The functional properties of a resource denote its ability to provide certain services. The nonfunctional properties mainly denote constraints, such as available memory and computational capacity. In this prototype, the user profile was not taken into account. When the composition is ready, the selected resources are advertised to the user using visual and audio feedback.

The second alternative, *semiautonomous composition*, is otherwise similar to the first one but the system provides an ordered list of resource combinations to the user and lets the user select one. The combinations are ordered based on the system's optimality criteria. A user can constrain the autonomous composition by determining some resources before the autonomous composition is performed.

Finally, in *manual composition*, a user selects all local resources that he wants to use. The system, however, is still in control of selecting the resources not present in the local environment, such as server repositories or databases. These resources can be selected based on a user profile. This manual method is the most interesting from the NFC point of view, since the user selects the resources by touching the RFID tags attached to them. The RFID tags are entity tags: each tag contains the ID and name of the resource. The same names are advertised to users in the icons. The RFID tags are located on the device or close to it. For those devices that are not accessible (for example, a projector hanging on the ceiling), a control board is located in some visible place. This control board has an icon for selecting each resource in the room that is not accessible or is not in the room, such as Internet servers.

When a user enters a room, she touches first the start tag of the desired service. The mobile phone screen informs the user of the resource types that must be selected before using the application. The GUI shows a stack with empty icons, each icon representing one resource type. The user selects a resource by touching an icon in the environment identifying that resource. When a tag is touched, the corresponding icon is filled on the GUI, and the name of the touched resource is shown as well. When the user has selected all resources, the application is started. The user can replace already-selected resources with new ones by just touching the new resources icon.

We have performed a usability test comparing the three methods. The results indicate that users prefer to use the manual method when they are somehow familiar with the environment and also in public places when an application involves private content such as personal photographs.

This application was used to test both the discovery and the composition stages described in our interaction model. The icons placed close to the resources that the user could select were of type Entity. Although the implementation described in [10] used other formats for the tag, the best approach is the one described here. The URI of the Entity type contains the place where the device is located, the type of resource (display, speaker), and the ID of that resource. A Text record type stores the type of resource as a text string, followed by a name identifying this resource from others ("display close to the big screen," for example). The parameters can also contain information that might be useful for a user when making a decision. For a display, for example, display resolution and network speed could be described.

Place Messaging

PlaceMessaging is a place-based message board service. This service provides a location-related message box having various user interfaces. Using this message box, people can share thoughts, opinions, greetings, announcements, and news. Message boards and guest books are just two examples of such message boxes. Message boards can be found in shopping malls, university halls, community centers, to name a few. Notes left on message boards can be related to the place, or the board can be just provided by the place and used freely. On the other hand, messages written in guest books are usually closely related to the place. Recently, guest books associated with web portals and blogs have become a part of our daily life. In another recent development, TV viewers can share comments about the TV shows they are watching. The messages sent are usually shown superimposed in the lower part of the TV screen.

PlaceMessaging allows users to post comments related to a place using their NFC phones. In the same way, users can also read messages posted by others. Messages can be presented to the users on the mobile phone's screen or on a wall display. Each board has an associated RFID tag that is placed behind an icon advertising the PlaceMessaging service. When a user touches such a tag with her mobile phone, the PlaceMessaging application is launched. The user is presented with a list of messages that have been dropped to that place. The user can read the messages and also write his or her own message; the message can also contain images. The user can then drop the new message to that place by touching the tag again. If the PlaceMessaging tag in question has a related wall display nearby, the user can read the messages from that display, and the message dropped by the user will be shown at that display. Figure 7.10 shows an example of the UI of place messaging.

In PlaceMessaging, "a place" is not necessarily a physical location. Place can be any topic one wants to gather messages about. It can

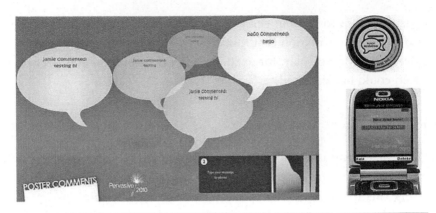

Figure 7.10   PlaceMessaging user interface.

refer to a physical place (e.g., to a restaurant), to something that is happening at a place (e.g., to a conference), or to a person in a place (e.g., to Mary at the office). The token is the object where the icon is placed. However, since each icon represents a place, the concrete place (room, restaurant, ...) could be considered the token. The record types used in this application are similar to the PickAndDrop types described before. Each tag identifies one place by a URL. That code is collected by a mobile phone each time the tag is touched, and it links the information picked and dropped by the user to the correct place information.

Interactive Poster

Interactive poster is an application that permits a poster presenter to provide an interactive experience to visitors. Interactive Poster offers a set of services that extend the possibilities of a classical poster presentation, making it more interesting to the audience. This application was initially designed for technical presentations in conferences, but can be easily adapted to other purposes such as marketing and product presentations. Interactive poster resembles the smart poster proposed by the NFC Forum. A smart poster service is triggered by touching an RFID tag on a physical poster with an NFC phone. The service presents multimedia content related to the poster on the phone's screen. This service can also be used to deliver other data to NFC phones.

Interactive Poster offers a wider set of services than the smart poster. The poster author can associate multimedia content with RFID tags placed on the poster (behind icons advertising the tags). Different icons on the poster give access to different content. Parameters stored in each tag determine a display and a playlist. Icons can be inserted as part of the poster during the poster design, or can be attached later. In both cases, special care is needed to design icons that do not disturb a visitor concentrating on the content presented on the physical poster. RFID tags can be either attached at the back side or inserted inside the poster material during the printing process.

When a user (visitor or author) touches one of these icons with an NFC phone, the corresponding service presents the content determined in the tag on the display determined in the tag. Each poster can have a wall display next to the poster, for example. The user can control the reproduction of the content using the mobile phone as a remote control. When the user presses a button in the mobile phone keypad, the corresponding command is sent to the display via the interactive space system.

Furthermore, Interactive Poster offers a set of services that can be used by the poster visitors. Those services are useful when the poster presenter is not at the poster location. Each service is advertised as an icon located in a control board that is placed close to the poster. Figure 7.11 shows this control board. When a user touches an icon, the associated service is started in the user's mobile phone. When multimedia content is required, it is shown in a nearby display. The services offered to the visitors by Interactive Poster are

- *Listen poster abstract.* The tag contains a record of type PickAndDrop. A URL identifies the audio file to be played. The audio file is played in the user's mobile phone's media player.
- *Give comments.* The comments are shown in a display nearby so that other visitors can read them. This service is based on the PlaceMessaging service explained earlier.
- *Pick multimedia content.* Visitors can pick the multimedia files that the author has provided and keep them in their personal storage. They can watch some of the files in their mobile phones, access them via a web application, and watch them

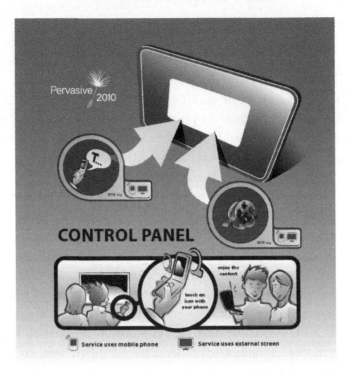

Figure 7.11   The control board for watching picked multimedia content on a wall display.

on a wall display located near the posters. The service is based on the Touch & Share service explained earlier.

- *Pick business card.* Visitors can pick the author's business card and store it in their personal repositories. Ideally, this tag should contain an Entity record describing the person, although for compatibility issues, the data is stored using the vcard standard.
- *Suggest a meeting.* Visitors can send SMSs (i.e., short messages with their NFC phones) to the author to suggest meetings. It uses the proprietary Nokia SMS format.

In a conference, fair, or similar event, each poster can be equipped with the Interactive Poster system. A display is located close to each poster for showing multimedia content. Furthermore, a separate wall display can be used to leave comments about the event. Visitors can also use this display to watch the multimedia content they have picked from the posters. A control board near the display contains icons for activating these services (Figure 7.11). This application was offered to the visitors of the Pervasive 2010 conference held in Helsinki, Finland, in May 2010.

In this case, the token is the poster itself.

Touch & Learn

A natural way to learn is by interacting with the environment. NFC technology allows, in the framework of interactive spaces described previously, augmenting everyday objects with information. Objects might be augmented, for example, with their names in several languages, description of the objects, or sounds related to the objects. The information depends on the learning goals. A space full of objects augmented with information forms a tangible learning environment.

With this concept in mind, we have created a series of applications called "Touch & Learn." Users carry NFC phones, and RFID tags are attached to objects in the environment. A learning application is started in a phone when a user touches a start tag. Touching objects with a phone, in turn, gives a command to the application or answers a question presented by the application, based on the information in the corresponding tag. Touch & Learn applications support a task-based learning method. The applications present tasks to the users, and have several modes determining the difficulty of tasks.

In general, Touch & Learn applications are divided into three main modes: Exploring, Practicing, and Mastering. Exploring is the least challenging mode. In this mode, a user explores her environment by picking information from different objects. Each time the user touches an augmented object, the mobile phone delivers the corresponding information to the user by showing text or images on the screen, playing an audio, or by performing a combination of these three. This way, a user starts learning in a natural way, by exploring her own environment. In other words, the task is just to explore the environment by touching objects with the phone.

In Practicing mode, the aim is that a user practices the previously acquired knowledge and skills. The tasks related to this mode always have the same pattern: the phone presents information, and the user has to find and touch the corresponding object in her environment. The user is given feedback indicating whether she touched the correct object. In Mastering mode, a user has to use the acquired and practiced information. Time constraints can be set, and multiplayer applications can be used as well. The task can be, for example, to fill missing information by touching objects.

In this application, the RFID tag contains one NDEF record of type Entity. At the same time, this record contains an array of NDEF records of type NFC Text that stores the name of the object or its representation in different languages. The tokens of our interaction model are the objects where the tag is placed.

Touch & Learn Languages   The first Touch & Learn application we implemented was the "Touch & Learn Languages." The aim of this application is to support users when learning new languages. In particular, Touch & Learn helps learning new vocabulary. We consider a natural way to approach a new vocabulary to be that the objects in the learner's environment tell you how they are named in the language being learnt.

We have observed that many people emulate this interaction by adding post-its to their everyday objects with the corresponding words. If we substitute the post-its with RFID tags, we get the same kind of interactive environment. Now a user just reads the name of an object from a phone's screen, after touching the object with the phone. The most obvious improvement is the usage of the audio modality: in addition to reading the word from the phone's screen, the user can listen to the word's pronunciation as well.

Moreover, as discussed earlier, different modes can be offered to users. In the exploring mode, the user just has to scan her environment looking for objects and then touch them. Each time the user touches an object, she hears the name of the object in the targeted language and can read the word from the phone's screen. Optionally, the phone can show an image representing the object as well. In the practicing mode, the phone presents words to the user. Each time the phone shows a word in the screen, and reads it aloud, the user has to touch the corresponding object. A sound-based feedback informs the user if the object touched is the correct one. For the mastering level, we have implemented a crossword game. The game is solved by selecting a row in the crossword and touching the object that corresponds to the word matching the row in question. The game can be played with or without a timer. If the user chooses to play with a timer, she will have to complete the game before time runs out.

Touch & Learn to Read   The second Touch & Learn prototype we have developed is called "Touch & Learn to Read." The purpose of this application is to support small children in kindergarten when they take their first steps in reading. In particular, the application supports the children in learning to read each other's names. This approach was selected as kindergartens usually have name tags identifying children's belongings; for example, each child has a nametag near his or her coat rack.

We placed RFID tags beneath the nametags. The children use the application by touching the nametags with NFC phones. A star icon on a nametag indicates the point to touch. An animation on the phones' screens presents a hand holding a phone and moving it to touch a star icon. The application has two different modes: Exploration and Practicing. The mode is selected by touching an animal icon representing that mode. These icons are placed on a control board that is attached on the wall of one of the kindergarten's rooms.

In the Exploration mode, the phone reads a name aloud and shows the name on its screen when a child touches a nametag with the phone. In the practice mode, the phone shows a name on the screen and waits for the child to touch the corresponding nametag. The phone also reads the name aloud periodically. Each time a nametag is touched, the phone informs the child with audio feedback if the answer was correct ("hienoa!"—which means "nice") or not ("yritä uudestaan"—which means "try again").

The most challenging task in the design of this application was to design the user interfaces for the targeted user group. As they are in their very first steps of learning to read, no textual feedback could be provided. The only text allowed is the actual names. We ended up with a very simple user interface. The animation and the names are the only content shown on the screen. No menus are provided, no keys need to be pressed—and if they are pressed, nothing happens. The application just reacts when a tag is touched.

This application was tested in a kindergarten in Oulu, Finland, in 2010. The children who used this Touch & Learn application were between 3 and 5 years old. We noticed that the children took to this technology naturally. The teachers had more difficulties in familiarizing themselves with the application, but the children rapidly took the phones and started exploring their environment.

A mastering mode for this application was being designed in Spring 2010. In the first version, children will have to look for a name that starts with a given letter (shown on the phone's screen). Further development will include complete spelling of the names. Moreover, we will attach RFID tags to other objects in the kindergarten. In this case, each icon represents a child and it is linked to the child by a "belonging relation."

Touch & Vote   This application is the virtual counterpart of casting a vote. Each candidate (a person, a presentation in a conference, etc.) has an icon. For example, in a poster session, a voting icon can be placed near each poster. Users can vote by touching an icon associated with the candidate they want to vote for. In a conference environment, visitors can vote for the best poster or the best demo, for example. In an exhibition, visitors can vote for the piece they like the most. Icons can be located either close to the physical objects or in a poster containing icons for all candidates. The last option is similar to the list of candidates in political elections, but instead of writing a candidate's number in a voting ballot, the visitor votes by touching the candidate's icon. This option is feasible, for example, to vote for a person as the employee of the month in a company or to vote for the best presentation in a conference.

The RFID tag contains two different record types. The first one is of type Entity and identifies the object or person that the user wants to vote for. The second is of type PickAndDrop and identifies the virtual ballot box where the user desires to vote. The process of voting is simple. The user selects the token he or she desires to vote, touches the associated icon with a mobile phone and, if required by the service, enters a personal ID. The token ID is sent to the server, which stores the vote of the user in the service database. Each category is associated with a different virtual ballot box. Depending on the server configuration, only one vote per category or several ones are accepted.

Here, the tokens represent two different entities: the object or person that the user desires to vote for and the ballot box or category where the user desires to insert this vote. This application was tested during the Pervasive 2010 conference held in Helsinki, Finland, in 2010.

Other Applications

Tourist Map Tourist Map is a map of the City of Oulu, Finland, enhanced with NFC technology. The most important places and attractions of the city (i.e., points of interest) are identified by icons on the map. RFID tags containing hyperlinks to web pages are placed behind the icons. Each web page URL contains one parameter identifying the point of interest. Each web page referred to by a URL contains links to three different web services. The first service offers information about the point of interest such as history, nearby hotels, and what to see there. The second service uses a local map web service to draw a path from the user's current position to the target place. Finally, the third service uses the local bus company's web service to show the next buses that go to the point of interest from the bus stop close to the user.

The Tourist Map has been designed to be used as an information board. A user stands in front of the map and studies the points of interests (POI). When a user touches a point of interest, a web page giving access to the three aforementioned services described is opened in the mobile phone's browser. The user can then read the general information, and study the route to the point of interest and the bus timetable as well. Making a call to a hotel near the POI is easy as well, when the phone number is listed in the general information. One main advantage of this application is that it uses already existing services to provide the information. Figure 7.12 shows a map of Oulu enhanced with NFC icons. In this example, the icon pictogram represents the type of place.

At the time of implementing this service, around 2006, this kind of services was rare, but in 2010 such map services are already being provided by companies, for example, the Smart Map provided by Red Solutions [43]. The difference between the Tourist Map and the Smart Map is that the Tourist Map has icons describing the type of the point of interest placed on the map, whereas the Smart Map presents standard NFC icons separately outside the map.

In this case, the token is the map, while each icon represents a place.

Locating Colleagues This application was implemented in 2006 and combines NFC technology with Ekahau's, an indoor location system based on Wi-Fi signal strength. The application can be used to

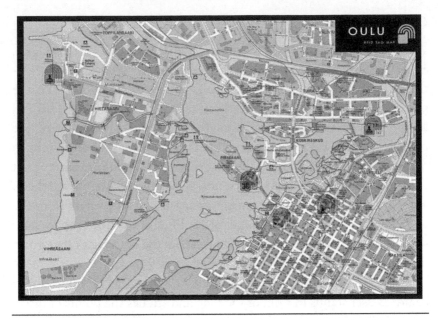

Figure 7.12   The Tourist Map.

locate a person or a mobile device. Each located resource is equipped with an Ekahau Wi-Fi tag. This tag reports its position regularly to the Ekahau positioning server. A resource can be located by touching that resource's icon on a control board. All located resource's icons can be collected on a single board, or individual icons can be attached to places where the location information is normally useful. The RFID tag contains a record of type Entity. When a resource icon is touched with an NFC phone, the phone requests a web page based on the URL stored in the NDEF record. When the web page is created, the location of the resource identified in the URL is requested from the positioning service, and a web page showing the position on a map is sent to the phone.

At a workplace, each staff member can carry a Wi-Fi tag. RFID tags with resource icons can be placed next to each staff member's office's doorway. The tags are entity tags storing staff members' IDs. When a person tries to meet a colleague but the colleague is not at his or her office, a single touch brings the location of the colleague on the person's mobile phone. We have implemented this scenario at our office environment.

Sheltered Home   This application, implemented in 2009, offers for a sheltered home a system to monitor the staff's visits to the elderly residents' apartments. Each house has two icons near the entrance door. Each tag contains an NDEF record of type State. The URL of that record identifies the house, while the Text identifies the state of the staff. One tag contains the state "in," while the other the state "out." When the sheltered home staff touches one of the icons, the state, the place ID, and the staff member ID are sent to the system that stores the information in a database. When a staff member enters an apartment, he or she touches an "in" tag with an NFC phone. When the staff member has performed all the needed tasks, he or she leaves. On leaving, he or she touches the "out," tag and a "leaving" event is sent to the server. The system replies by presenting a list of tasks on the mobile phone's display. The staff member checks the tasks that he or she just performed. As a result, the server has a log of visits to the apartments: which staff member visited, which apartment and when, how long the visit lasted, and which tasks were performed.

In this case, the token is the elderly house itself, and each tag represents the state of the staff member who assists this house.

Touch & Run   This application, implemented in 2007, is a board game that is distributed in the environment, for example, in a park [44]. Each square in a board is represented by a square icon in the environment (again, RFID tags are placed under the icons). Performing an action at a square requires touching the corresponding icon first with an NFC phone. The state of the game board is shown on the NFC phones' screens. Teammates can synchronize two phones by bringing them into contact and also by touching a synchronizing tag in the environment.

In the game we implemented, the goal is to occupy the whole board as quickly as possible. A single square is occupied by first reserving the four neighbors (north, east, south, and west) and then touching the tag in question. Reserving is performed by touching a tag; an occupied tag is interpreted as reserved. The status of each square is presented with colors on a mobile phone's screen. In addition, each square stays reserved for only a certain period of time (but an occupied cell stays occupied until the end of the game). As moving from one square to another means running in the environment, it is essential to minimize the overall distance. We performed some tests with this application in 2007. The game concept received positive feedback and the results suggested that spending some time planning between team members improved the performance.

A game board we have used for demonstrating the game is shown in Figure 7.13. The board visualizes a configuration in which the square icons are placed on a small island at downtown Oulu. The board has 16 squares. A synchronizing tag is placed at the top-right corner.

Cleaning Hospitals   This small field trial, performed in 2005, was one of our first NFC-related projects. We equipped an operation room of a local private hospital with just two icons, "Start" and "Stop." The cleaning staff touched a "Start" tag with an NFC phone when they started cleaning the room and "Stop" when they stopped. The tags contained SMS messages that were sent to their superior. She calculated the cleaning times based on the timestamps of the SMS messages. This simple test illustrated the potential of NFC: a very simple setup and just two touches helped to get information directly from a task critical to the overall hospital processes. In fact, the test revealed

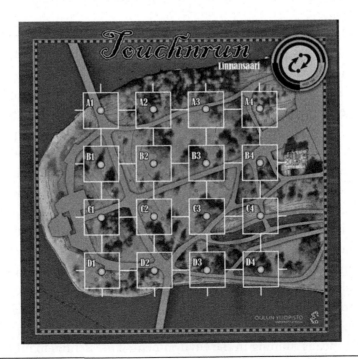

Figure 7.13   A game board for demonstrating Touch & Run.

that cleaning requires less time than was assumed, and hence the break reserved for cleaning was longer than necessary.

Discussion and Future Work

We described in this chapter how we are utilizing NFC technology in building user interfaces for interactive spaces. When compared to the presented types of applications publicly discussed as the first business cases (transit and ticketing, payment, and advertising), our prototypes offer much more functionality and richer user interfaces. When compared to other research in this area, similar work emphasizing icons for advertising RFID tags is rare.

We presented first the concept of interactive spaces: how these environments offer a rich set of services via natural user interfaces. We presented an interaction model for interactive spaces: how a user controls services by handling tokens. Moreover, we described how this interaction model can be realized by using NFC technology. The rest of the chapter described applications in which we have utilized NFC technology. Building these applications have been a valuable exercise for us; we have been able to evaluate different technological solutions and user interfaces in real usage. The concept of interactive spaces and the interaction model have been gradually developed during this development work.

As we stated earlier, the interaction model is not yet complete. Moreover, as we have worked from the applications to the model, the interaction in the presented applications was not designed based on the interaction model. Instead, we are using the implemented applications to verify our model: if we can describe the interaction of all these application using this model, we can state that the model is sufficient in describing these types of applications.

Our work will continue by building more applications and user interfaces. We will have more icons for sending commands to services and objects that can be used to control several services. We will also improve the REACHeS system described earlier. Other modalities such as haptics offer interesting possibilities when used together with NFC technology. An essential task will be to specify a language formalizing physical interaction in interactive spaces. The final goal is to have a language for defining the user interfaces and a platform for configuring the defined user interfaces and connecting them to services. All this will be performed dynamically: user interfaces will be created, modified, and dismantled as the situation requires when users move in their daily environment and use services based on their changing needs. We foresee that NFC technology will play a central role in achieving this goal.

References

1. Weiser, M. (1991). The computer for the twenty-first century. *Scientific American*, 94–100, September.
2. Greenfield, A. (2006). *Everyware: The Dawning Age of Ubiquitous Computing*. Peachpit Press, Sebastopol, CA.
3. Dey, A.K. and Abowd, G.D. (2000). Towards a better understanding of context and context-awareness. *Workshop on the What, Who, Where, When and How of Context Awareness, Affiliated with the 2000 ACM Conference on Human Factors in Computer Systems (CHI 2000)*.
4. Dahl, Y. (2008). Modeling human–computer interaction in smart spaces: Existing and emerging techniques. In Asai, K. (Ed.), *Human Computer Interaction: New Developments* (pp. 177–190). InTech, Rijeka, Croatia.
5. Valli, A. Natural Interaction White Paper. Downloaded from http://naturalinteraction.org/images/whitepaper.pdf (last accessed 18.06.2010).
6. Sanchez, I., Riekki, J., and Pyykkönen, M. (2009). Touch & Compose: Physical user interface for application composition in smart environments. In *Proceedings of the 1st International Workshop on Near Field Communication—NFC'09*. Hagenberg, Austria.
7. Skål webpage: http://www.skaal.no/ (last accessed on 18.06.2010).
8. Want, R., Fishkin, K. P., Gujar, A., and Harrison, B.L. (1999). Bridging physical and virtual worlds with electronic tags. In *Proceedings of the SIGCHI Conference on Human Factors in Computing Systems*, 370–377, Pittsburgh, PA.
9. Ishii, H. and Ullmer, B. (1997). Tangible bits: Towards seamless interfaces between people, bits and atoms. In *Proceedings of SIGCHI Conference on Human Factors in Computing Systems*, 234–241, Atlanta, GA.
10. Davidyuk, O., Sánchez, I., and Riekki, J. CADEAU: Supporting autonomic and user-controlled application composition in ubiquitous environments. In M. Apostolos (Ed.), *Pervasive Computing and Communications Design and Deployment: Technologies, Trends, and Applications*. IGI Global. To be published.
11. Arnall T. (2006). A graphic language for touch-based interactions. In *Proceedings of Mobile Interaction with the Real World (MIRW) in Conjunction with the 8th International Conference on Human Computer Interaction with Mobile Devices and Services (MobileHCI 2006)*, 18–22, Espoo, Finland.

12. Tungare, M., Pyla, P.S., Bafna, P., Glina, V., Zheng, W., Yu, X., Balli, U., and Harrison, S. (2006). Embodied data objects: Tangible interfaces to information appliances. In *Proceedings of the 44th ACM Southeast Conference (ACM SE'06)*, March 10–12, Melbourne, Florida, pp. 359–364.

13. Välkkynen, P., Tuomisto, T., and Korhonen, I. (2006). Suggestions for visualising physical hyperlinks, *Proceedings of the Pervasive Mobile Interaction Devices*, May, Dublin, Ireland, pp. 245–254.

14. Riekki, J., Salminen, T., and Alakarppa, I. (2006). Requesting pervasive services by touching RFID tags, *IEEE Pervasive Computing*, pp. 40–46, January–March, 2006.

15. Riekki, J., Sanchez, I., and Pyykkönen, M. (2010). Remote control for pervasive services. *International Journal of Autonomous and Adaptive Communications Systems*, 3(1):39–58.

16. Broll, G., Keck, S., Holleis, P., and Butz, A. (2009). Improving the accessibility of NFC/RFID-based mobile interaction through learnability and guidance. In *Proceedings of the 11th International Conference on Human–Computer Interaction with Mobile Devices and Services (MobileHCI '09)*, September 15–18, Bonn, Germany, 2009.

17. Van der Lelie, C. (2006). The value of storyboards in the product design process. *Personal and Ubiquitous Computing*, 10(2):159–162.

18. Davidoff, S., Lee, M.K., Dey, A., and Zimmerman, J. (2007). Rapidly exploring application design through speed dating. In *Proceedings of the 9th International Conference on Ubiquitous Computing*. Innsburck, Austria.

19. Diekmann, T., Melski, A., and Schumann, M. (2007). Data-on-network vs. data-on-tag: Managing data in complex RFID environments. *Proceedings of the 40th Annual Hawaii International Conference on System Sciences (HICSS'07)*, p. 224a, , 2007.

20. Riekki, J., Sanchez, I., and Pyykkönen, M. (2009). Universal remote control for the smart world. In *Proceedings of 5th International Conference on Ubiquitous Intelligence and Computing (UIC 2008)*, p. 563–577, Oslo, Norway.

21. Sánchez, I., Kuusela, E., Turpeinen, S., Röning, J., and Riekki, J. 2009. Botnet-inspired architecture for interactive spaces. In *Proceedings of the 8th international Conference on Mobile and Ubiquitous Multimedia (MUM '09)*, November 22–25, Cambridge, U.K., 2009.

22. Sanchez, I., Riekki, J., and Pyykkönen, M. (2008), Touch & Control: Interacting with services by touching RFID tags. In *Proceedings of the Second International Workshop on RFID Technology (IWRT 2008)—Concepts, Applications, Challenges*, in conjunction with ICEIS 2008, Barcelona, Spain, June 12–13, pp. 53–62; Available at: http://www.ee.oulu.fi/research/isg/publications/ID/1272.

23. Turunen, M., Kallinen, A., Sànchez, I., Riekki, J., Hella, J., Olsson, T., Melto, A., Rajaniemi, J., Hakulinen, J., Mäkinen, E., Valkama, P., Miettinen, T., Pyykkönen, M., Saloranta, T., Gilman, E., and Raisamo, R. (2009). Multimodal interaction with speech and physical touch interface

in a media center application. In *Proceedings of the International Conference on Advances in Computer Entertainment Technology (ACE '09)*, October 29–31, Athens, Greece, 2009.

24. Sánchez, I., Riekki, J., Rousu, J., and Pirttikangas, S. (2008). Touch & Share: RFID based ubiquitous file containers. In *Proceedings of the 7th International Conference on Mobile and Ubiquitous Multimedia (MUM '08)*, December 03–05, Umeå, Sweden, 2008.

25. Hosio, S., Kawsar, F., Riekki, J., and Nakajima, T. DroPicks: A tool for collaborative content sharing exploiting everyday artifacts. In *Proceedings of 4th International Symposium on Ubiquitous Computing Systems (UCS07)*, Tokyo, Japan (pp. 258–265), 2007.

26. Rousu, J. Virtual file repository for mobile phones. Master's thesis. Department of Electrical and Information Engineering, University of Oulu. May 2008. Available at: http://www.ee.oulu.fi/research/isg/publications/ID/1209 (last accessed 12-06-2010).

27. Hornecker, E. (2009). Encyclopedia entry on tangible interaction. Retrieved June 30, 2010 from Interaction-Design.org: http://www.interaction-design.org/encyclopedia/tangible_interaction.html.

28. Holmquist, L., Redström, J., and Ljungstrand, P. (1999). Token-based access to digital information. *Lecture Notes on Computer Science 1707*, Springer, 1999, pp. 234–245.

29. Abowd, G.D. (1999). Classroom 2000: An experiment with the instrumentation of a living educational environment. *IBM Systems Journal* 38(4):508–530. Special issue on Pervasive Computing.

30. Johanson, B., Fox, A., and Winograd, T. (2002) The interactive workspaces project: Experiences with ubiquitous computing rooms. *IEEE Pervasive Computing*, 1(2): 67–74, April 2002.

31. Pering, T., Ballagas, R., and Want, R. 2005. Spontaneous marriages of mobile devices and interactive spaces. *Communications of the ACM* 48(9): 53–59, September 2005.

32. International Telecommunications Union (ITU) (2005). The Internet of things. Executive Summary. Retrieved June 30, 2010 from http://www.itu.int/dms_pub/itu-s/opb/pol/S-POL-IR.IT-2005-SUM-PDF-E.pdf.

33. Ferscha A., Hechinger M., Mayrhofer R., dos Santos Rocha M., Franz M., and Oberhauser, R. (2004). Digital Aura. In *Proceedings of Advances in Pervasive Computing*, Video paper at Pervasive 2004 conference, Vienna, Austria, April 18–23, 2004.

34. Ullmer, B. and Ishii, H. (2000). Emerging frameworks for tangible user interfaces. *IBM Systems Journal* 39(3–4): 915–931.

35. Koleva, B., Benford, S., Hui Ng K., and Rodden, T. (2003). A framework for tangible user interfaces. In *Proceedings of the Real World User Interfaces Workshop at the 5th International Symposium on Human–Computer Interaction with Mobile Devices and Services (MobileHCI 2003)*, Udine, Italy, September 2003.

36. Hornecker, E. and Buur, J. (2006). Getting a grip on tangible interaction: A framework on physical space and social interaction. *Proceedings of the SIGCHI Conference on Human Factors in Computing Systems*, April 22–27, 2006, Montréal, Québec, Canada.

37. Norman, D.A. (1999). Affordance, conventions, and design. *Interactions* 6(3): 38–43, May 1999.

38. P. Saint-Andre, Ed. (2004). Extensible Messaging and Presence Protocol (XMPP): Core. IETF RFC 3920.

39. Soininen, J.P., Liuha, P., Lappeteläinen, A., Honkola, J., Främling, K., and Raisamo, R. (2010). Device interoperability: Emergence of the smart environment ecosystems White Paper accessible from http://www.tivit.fi/fi/dokumentit/64/DIEM%20whitepaper.pdf. Last accessed 01.07.2010.

40. Richardshon, L. and Ruby, S. (2007). RESTful Web Services. Editorial O'Reilly.

41. NFC Forum (2010). NFC Data Exchange Format, NDEF. Retrieved June 8, 2010, from http://www.nfc-forum.org/specs/.

42. NFC Forum (2010). NFC Record Type Definition (RTD) Technical Specification. Retrieved June 8, 2010, from http://www.nfc-forum.org/specs/.

43. Accessible from http://www.redsolution.fi/applications_1/smartmap.html. Retrieved June 30, 2010.

44. Riekki, J., Sasin, S., and Pirttikangas, S. (2008). Touchnrun: An RFID-based distributed board game motivating to move. *The Third International Conference on Persuasive Technology (Persuasive 2008)*, June 4–6, Oulu, Finland. Poster Proceedings, University of Oulu, Department of Information Processing Science, A43, pp. 34–40.

8

Privacy-Preserving Receipt Management with NFC Phones

NING SHANG, ELISA BERTINO, AND KEVIN STEUER JR.

Contents

The combined use of the Internet and mobile technologies is leading to major changes in how individuals communicate, conduct business transactions, and access resources and services. In such a scenario, digital identity management technology is fundamental for enabling transactions and interactions across the Internet. In this article, we demonstrate an application for the privacy-preserving management of users' identity attributes on Near Field Communication (NFC) mobile devices.*

Introduction

Users increasingly use mobile devices to communicate, conduct business transactions, and access resources and services. In such a scenario, digital identity management technology is fundamental in customizing user experience, protecting privacy, underpinning accountability in business transactions, and in complying with regulatory controls. Digital identity consists of data, referred to as *identity attributes*, that encode properties of individuals. Such properties are often the basis on which access to sensitive resources and services is given. Because they may encode sensitive information, identity attributes can be the target

---

* This article is based on the IDtrust 2009 publication "Privacy-preserving management of transactions' receipts for mobile environments" [13].

of attacks: the loss or theft of mobile devices resulting in an exposure of identity attributes; identity attributes being sent over Wi-Fi or 3G/4G networks and easily intercepted; identity attributes being captured via Bluetooth connections without the user's consent; and mobile viruses, worms, and Trojan horses accessing the identity attributes stored on mobile devices if this information is not properly protected. Therefore, assuring privacy and security of identity attributes stored on mobile devices is crucial. We propose to demonstrate a system for the privacy-preserving management of users' identity attributes in the form of electronic transaction receipts on mobile devices. The system is based on the concept of privacy-preserving multifactor verification of identity attributes achieved using a set of cryptographic protocols.

NFC is a short-range high-frequency wireless connectivity technology that provides *intuitive, simple, and safe communication between electronic devices* [10]. Two NFC-compatible devices communicate (exchange data) over a distance of a few centimeters. An NFC communication can be initialized by simply "touching" or "tapping" one NFC device onto another. Such a short transmission range makes eavesdropping inherently hard, and thus helps achieve better communication security. Suitable to be used with a variety of devices, NFC technology is mainly aimed at mobile phones. The computational power of modern mobile phones and NFC wireless connectivity technology have become the enabler of various mobile-based applications, such as electronic ticketing for public transportations, mobile payment, and bootstrapping Wi-Fi or Bluetooth pairing processes.

We illustrate an application in the context of a mobile commerce scenario in which electronic transaction receipts are used by customers and service providers (stores) to facilitate further transactions. The combined use of NFC and cryptographic technologies makes such transactions simple, fast, and privacy preserving.

Receipt Management Overview

Establishing mutual trust between customers, users and service providers is critical. A possible approach to establish trust is to view the transactions users have carried out in the past. The history of former transactions gives information about users' behavior, their abilities, and their dispositions, and thus helps to decide whom to trust. Yahoo! Auction, Amazon, and eBay are examples of systems that rate both users and service providers, based on their past interactions history. Maintaining the history of user transactions and establishing trust based on these transactions and other factors is a complex task. An important component of any such solution is represented by systems managing receipts of transactions. By receipts, we refer to information that characterizes a transaction, such as the amount paid and the service provider for the transaction.

Managing transaction receipts on mobile devices is very challenging. On one hand, the sharing of information about transactions should be facilitated among service providers. A customer should be able to disclose to a service provider a view of his or her past transactions with other service providers in order to get discounts or to prove good behavior over the past. On the other hand, transaction receipts need to be protected as they may convey sensitive information about a user and can be the target of attacks. Moreover, users should be able to control which service provider has access to information about their past interactions. Assuring privacy and security of transactions' receipts, as well as of any sensitive information, in the context of mobile environments is further complicated by the fact that mobile devices are not secure. Recent statistics [1] show that millions of lost or stolen mobile devices that store users' sensitive data have been reported. In addition to loss or theft, there are an increasing number of viruses, worms, and Trojan horses targeting mobile devices. Moreover, current attacks against Bluetooth and well-known WLAN and GPRS vulnerabilities show that it is very easy for attackers to compromise mobile devices [18]. Another issue is related to how service providers determine whether users are trusted, based on their past transactions. Trust establishment should be a policy-driven process. Service providers should specify policies stating the conditions that a user's transaction receipts must satisfy for the user to be trusted or to get service with favorable conditions. An example of such a policy is that a user can receive a discount if he or she has spent $50 or more. Thus, an important requirement is the introduction of policy language that allows service providers to express conditions against transaction receipts.

To address such issues, we propose a policy-based approach for the management of a user's transaction history on mobile devices that provides:

1. Integrity, confidentiality, and privacy of users' transaction information
2. Selective and minimal disclosure of transaction information
3. Trust establishment based on transaction history

Our approach allows a user to prove to a service provider that he or she has performed a transaction satisfying a set of conditions by such a service provider without revealing any other information about the transaction. The approach is based on the notion of *transaction receipts* issued by service providers upon a successful transaction. Our approach combines a zero-knowledge proof of knowledge (ZKPK) technique [15] and Oblivious Commitment-Based Envelope (OCBE) protocols [9] to assure privacy of information recorded in the receipts.

Background

In this section, we introduce the basic cryptographic notions on which our transaction receipts management approach is based.

Discrete Logarithm Problem and Computational Diffie–Hellman Problem

Let G be a multiplicatively written cyclic group of order q, and let g be a generator of G. The map $\varphi: Z\ TG$, $\varphi(n) = g^n$ is a group homomorphism with kernel $\mathbb{Z}_q$. The problem of computing the inverse map of φ is called the *discrete logarithm problem (DLP) to the base of g*.

For a multiplicatively written cyclic group G of order q, with a generator $g \in G$, the *computational Diffie–Hellman problem* is the following: Given g^a and g^b for randomly-chosen secret $a, b \in \{0,\dots, q-1\}$, compute g^{ab}.

Pedersen Commitment

The Pedersen commitment scheme, first introduced in Reference 14, is an unconditionally hiding and computationally binding commitment scheme that is based on the intractability of the DLP. The scheme is originally described with a specific implementation that uses a subgroup of the multiplicative group of a finite field. Note that this choice of implementation is not intrinsic to the Pedersen commitment scheme itself; it can be implemented with any suitable abelian groups, for example, elliptic curves over finite fields. Therefore, we rewrite the Pedersen commitment scheme in a more general language as follows.

Setup A trusted third party T chooses a finite cyclic group G of large prime order p so that the computational Diffie–Hellman problem is hard in G. Write the group operation in G as multiplication. T chooses an element $g \in G$ as a generator, and another element $h \in G$ such that it is hard to find the discrete logarithm of h with respect to g, that is, an integer α such that $h = g^\alpha$. T may or may not know the number α. T publishes G, p, g and h as the system's parameters.

Commit The domain of committed values is the finite field $\mathbb{F}_p$ of p elements, which can be represented as the set of integers $\mathbb{F}_p = \{0, 1, \ldots, p-1\}$. For a party U to commit a value $x \in \mathbb{F}_p$, it randomly chooses $r \in \mathbb{F}_p$, and computes the commitment $c = g^x h^r \in G$.

Open U shows the values x and r to open a commitment c. The verifier checks whether $c = g^x h^r$.

ZKPK Protocol

It turns out that in the Pedersen commitment scheme described above, a party U referred to as the *prover*, can convince the verifier, V, that U can open a commitment $c = g^x h^r$, without showing the values x and r in clear. Indeed, by following the ZKPK protocol below, V will learn nothing about the actual values of x and r. This ZKPK protocol, which works for Pedersen commitments, is an adapted version of the zero-knowledge proof protocol proposed by Schnorr [15].

ZKPK (Schnorr Protocol) As in the case of the Pedersen commitment scheme, a trusted party T generates public parameters G, p, g, and h. A prover U who holds private knowledge of values x and r can convince a verifier V that U can open the Pedersen commitment $c = g^x h^r$ as follows.

1. U randomly chooses $y, s \in \mathbb{F}_p^*$, and sends V the element $d = g^y h^s \in G$.
2. V picks a random value $e \in \mathbb{F}_p^*$, and sends e as a challenge to U.
3. U sends $u = y + ex$, $v = s + er$, both in $\mathbb{F}_p$, to V.
4. V accepts the proof if and only if $g^u h^v = d \cdot c^e$ in G.

Shamir's Secret Sharing Scheme

Shamir's (k, n) threshold scheme [16] is a method that divides a secret into n shares and allows the secret to be reconstructed if and only if any k shares are present. Here k and n are both positive integers and $k \Psi n$. It is also called *Shamir's secret sharing scheme*.

The scheme works as follows. A trusted party, T, chooses a finite field $\mathbb{F}_p$ of p elements, with p large enough. Let the secret message S be encoded as an element $a_0 \in \mathbb{F}_p$. T randomly chooses $k - 1$ elements $a_1, \ldots, a_{k-1} \in \mathbb{F}_p$, and constructs a degree $k-1$ polynomial $f(x) = a_0 + a_1 x + \cdots + a_{k-1} x^{k-1} \in \mathbb{F}_p$. T chooses n elements $\alpha_1, \alpha_2, \ldots, \alpha_n \in \mathbb{F}_p$, and creates the secret shares S_i as pairs

$$S_i = (\alpha_i, f(\alpha_i)), 1 \leq i \leq n,$$

where $f(\alpha_i)$ is the polynomial evaluation of f at α_i. Given any subset of k such shares, the polynomial $f(x)$, of degree $k - 1$, can be efficiently reconstructed via interpolation (see, e.g., Reference 3, Section 2.2). The secret S, encoded as the constant coefficient a_0, is thus recovered.

Shamir's (k, n) threshold scheme has many good properties. Most prominently, it is information theoretically secure, in the sense that the knowledge of less than k shares gives no information about the secret S better than guessing, and it is minimal, in that the size of each share does not exceed the size of the secret. Interested readers can refer to Reference 16 for more details.

OCBE Protocols

The OCBE protocols, proposed in Reference 9, provide the capability of enforcing access control policies in an oblivious way. Three communications parties are involved in OCBE protocols: a receiver Re, a sender Se, and a trusted third party T. More precisely, the OCBE protocols ensure that the receiver Re can decrypt a message sent by Se if and only if its committed value satisfies a condition given by a predicate in Se's access control policy, while Se learns nothing about the committed value. The possible predicates are comparison predicates $=$, $\neq$, $>$, $\Box$, $<$, and Y.

The OCBE protocols are built with several cryptographic components:

1. The Pedersen commitment scheme.
2. A semantically secure symmetric-key encryption algorithm E, for example, AES, with key length k-bits. Let $E_{key}[M]$ denote the encrypted message M under the encryption algorithm E with symmetric encryption key.
3. A cryptographic hash function $H(\cdot):\{0,1\}^*T\{0,1\}^k$. When we write $H(\alpha)$ for an input α in a certain set, we adopt the convention that there is a canonical encoding which encodes α as a bit string, that is, an element in $\{0,1\}^*$, without explicitly specifying the encoding.

Given the notation as above, we summarize the EQ-OCBE and GE-OCBE protocols; that is, the OCBE protocols for $=$ and $\Box$ predicates, respectively, in what follows. The OCBE protocols for other predicates can be derived and described in a similar fashion. The protocols are stated in a slightly different way than in [9], to better suit the presentation in this paper.

EQ-OCBE Protocol   Parameter generation T runs a Pedersen commitment setup protocol to generate system parameters $Param = ``G, g, h \geq$. T also outputs the order of G, p, and $\mathcal{P} = \{EQ_{x_0} : x_0 \in \mathbb{F}_p\}$, where

$$EQ_{a_0} : \mathbb{F}_p \to \{true, false\}$$

is an equality predicate such that $EQ_{x_0}(x)$ is true if and only if $x = x_0$.

Commitment   T first chooses an element $x \in \mathbb{F}_p$ for Re to commit. T then randomly chooses $r \in \mathbb{F}_p$, and computes the Pedersen commitment $c = g^x h^r$. T sends x, r, c to Re, and sends c to Se.*

Interaction

- Re makes a data service request to Se.
- Based on this request, Se sends an equality predicate $EQ_{x_0} \in \mathcal{P}$.
- Upon receiving this predicate, Re sends a Pedersen commitment $c = g^x h^r$ to Se.
- Se randomly picks $y \in \mathbb{F}_p^*$, computes $\sigma = (cg^{-x_0})^y$, and sends to Re a pair $\langle \eta = h^y, C = E_{H(\sigma)}[M] \rangle$, where M is the message containing the requested data.

Open   Upon receiving $\langle \eta, C \rangle$ from Se, Re computes $\sigma' = h^r$, and decrypts C using $H(\sigma')$.

GE-OCBE Protocol

Parameter generation   As in EQ-OCBE, T runs a Pedersen commitment setup protocol to generate system parameters $Param = \langle G, g, h \rangle$, and outputs the order of G, p. In addition, T chooses another parameter ,, which specifies an upper bound for the length of attribute values, such that $2^, < p/2$. T also outputs $V = \{0, 1, \dots, 2^,-1\} \cup \mathbb{F}_p$, and $\mathcal{P} = \{GE_{x_0} : x_0 \in V\}$, where

$$GE_{x_0} : V \to \{true, false\}$$

is a predicate such that $GE_{x_0}(x)$ is true if and only if $x \square x_0$.

Commitment   This step is the same as EQ-OCBE. T chooses an integer $x \in V$ for Re to commit. T then randomly chooses $r \in \mathbb{F}_p$, and computes the Pedersen commitment $c = g^x h^r$. T sends x, r, c to Re, and sends c to Se.†

---

* In an off-line alternative, T can digitally sign c and sends x,r,c and the signature of c to Re. Then the validity of the commitment c can be ensured by verifying T's signature. In this way, after Se obtains T's public key for signature verification, no communication is needed between T and Se.

† Similarly, an off-line alternative also works here.

Interaction

- Re makes a data service request to Se.
- Based on the request, Se sends to Re a predicate $GE_{x_0} \in \mathcal{P}$.
- Upon receiving this predicate, Re sends to Se a Pedersen commitment $c = g^x h^r$.
- Let $d = (x - x_0) \pmod p$. Re picks $r_1, \ldots, r_{-1} \in \mathbb{F}_p$, and sets $r_0 = r - \sum_{i=1}^{\ell-1} 2^i r_i$. If $GE_{x_0}(x)$ is true, let $d_{-1} \ldots d_1 d_0$ be d's binary representation, with d_0 the lowest bit. Otherwise if GE_{x_0} is false, Re randomly chooses $d_{-1}, \ldots, d_1 \in \{0,1\}$, and sets $d_0 = d - \sum_{i=1}^{\ell-1} 2^i d_i \pmod p$. Re computes , commitments $c_i = g^{d_i} h^{r_i}$ for $0 \Psi\ i\ \Psi\ ,-1$, and sends all of them to Se.
- Se checks that $cg^{-x_0} = \prod_{i=0}^{\ell-1}(c_i)^{2^i}$. Se randomly chooses , bit strings $k_0, \ldots, k_{-1}$, and sets $k = H(k_0 ||\ldots|| k_{-1})$. Se picks $y \in \mathbb{F}_p^*$, and computes $\eta = h^y$, $C = \varepsilon_k[M]$, where M is the message containing requested data. For each $0 \Psi\ i\ \Psi\ ,-1$ and $j = 0,1$, Se computes $\sigma_i^j = (c_i g^{-j})^y, C_i^j = H(\sigma_i^j)\ k_i$. Se sends to Re the tuple

$$\left\langle \eta, C_0^0, C_0^1, \ldots, C_{\ell-1}^0, C_{\ell-1}^1, C \right\rangle.$$

Open   After Re receives the tuple $\left\langle \eta, C_0^0, C_0^1, \ldots, C_{\ell-1}^0, C_{\ell-1}^1, C \right\rangle$ from Se as above, Re computes $\sigma_i' = \eta^{r_i}$, and $k_i' = H(\sigma_i')\ C_i^{d_i}$, for $0 \Psi\ i\ \Psi\ ,-1$. Re then computes $k' = H(k_0' ||\ldots|| k_{-1}')$, and decrypts C using key k'.

LE-OCBE, the OCBE protocol for the Ψ predicates, can be constructed in a similar way as GE-OCBE. Other OCBE protocols (for $\neq$, $<$, $>$ predicates) can be built on EQ-OCBE, GE-OCBE and LE-OCBE.

All these OCBE protocols guarantee that the receiver Re can decrypt the message sent by Se if and only if the corresponding predicate is evaluated as true at Re's committed value, and that Se does not learn anything about this committed value.

Note that for certain applications, we can let Se know whether Re's committed value satisfies the specified predicate, by extending the OCBE protocols with one more step: Re shows to Se the decrypted message.

Nokia 6131 NFC phone

The device we use to illustrate our receipt management protocols is a Nokia 6131 NFC [12] mobile phone.

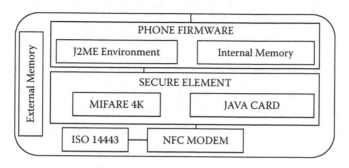

Figuro 8.1   Nokio 6131 NFC.

The Nokia 6131 NFC phone architecture is shown in Figure 8.1. It consists of an Antenna, for detecting external targets such as tags, external readers, or other Nokia 6131 NFC mobile phones; an NFC modem, for providing the capability to send and receive commands between antenna, secure element and phone firmware including J2ME environment; a Secure element, for enabling third-party application development using tag/card emulation; Phone firmware, for providing mobile phone functions with NFC features; a SIM card, for GSM subscription identification and service management; J2ME environment included in phone firmware, for enabling third-party application development using Nokia 6131 NFC features; and External memory.

The Secure element within Nokia 6131 NFC can store information securely, which can be used for payment and ticketing applications or for access control and electronic identifications. The Secure element is divided into two subcomponents, a Java Card area (also referred to as smart card) and a Mifare 4K area. The Mifare 4K area can be considered as a memory with access control, and typically it is simpler to implement than a smart card application. The Mifare 4K contains data, whereas smart card application contains an executable program. The Java Card provides high security environment and executes code, which means it can be used for more complex applications. The Secure element is accessible through NFC modem internally from MIDlets and externally by external readers. MIDlets are Java applications running in the J2ME environment.

Receipt Management System

Receipt Format

A service provider, upon the completion of a transaction, usually sends the user a receipt that specifies a set of information about the transaction such as user identifier, the identifier of the service provider, the purchased items, the price paid for the items, the quantity, the date of the transaction, and shipment and billing information. We denote this type of information as *transaction attributes*.

We consider only a subset of the possible attributes that can be associated with a transaction. The subset includes the user identifier, the service provider identifier, the category to which the purchased item belongs to, the price of the item, and the transaction date, because they are the more relevant attributes to establish trust in the user.

We assume that service providers have a public key infrastructure that allows them to issue users signed transaction receipts. In particular, we assume that each service provider is associated with a pair of keys $(K_{\mathrm{Priv}}, K_{\mathrm{Pub}})$ where K_{Priv} is the private key used to sign the transaction receipts and K_{Pub} is the public key used by other service providers to verify authenticity and integrity of receipts. In order to support a privacy-preserving proof of the possession of such receipts, the transaction receipts released under our protocol include the transactions' attributes and their corresponding Pedersen commitments. The Pedersen commitments of a transaction attributes are used by a user to prove the possession of the receipt of this transaction to other service providers. To compute the Pedersen commitments of the transaction attributes, the service provider runs the Pedersen commitment setup protocol described in the section titled "Pedersen Commitment" to generate the parameters $\mathrm{Param} = \langle G, g, h \rangle$. Then, the service provider publishes $G, p, g,$ and h and its public key K_{Pub}. The structure of transaction receipts is defined as follows.

Transaction Receipt   Let SP be a service provider and U be a user with which SP has successfully carried out a transaction, Tr. Let $(G, p, g, h,$ and $K_{\mathrm{pub}})$ be the public parameters of SP. The receipt for transaction Tr carried out by U and SP has the following format:

TRAN-ID	ATTR	COM	SIG

TRAN-ID	ATTR			COM		SIG
1234	BUYER	John Smith		BUYER	7645353 6366363	1124457 6590873 3647688
	SELLER	BookStore.com		SELLER	1312425 54546	
	CATEGORY	Books		CATEGORY	2224223 525	
	PRICE	30		PRICE	1341515	
	DATE	11-04-2008		DATE	1315657	

Figure 8.2 A sample receipt.

where *TRAN-ID* is the transaction identifier; *ATTR* is the set of transaction attributes {*BUYER, SELLER, CATEGORY, PRICE, DATE*} where

1. *BUYER* is the user identifier.
2. *SELLER* is the service provider's identifier.
3. *CATEGORY* is the selling category of the item being bought.
4. *PRICE* is the price of the item.
5. *DATE* is the date of the transaction.

COM is the set of the Pedersen commitments of the attributes in *ATTR*. Each element in *COM* is a tuple of the form $\langle A, COMMIT \rangle$ where A is the value of an attribute in *ATTR*, *COMMIT* is the Pedersen commitment $g^A h^r$ of A, and r is a random secret known only to U. *SIG* is the signature of service provider *SP* on *COM*. In what follows, we will use the dot notation to denote the different components of transaction receipt.

Example 4.1. Suppose that John Smith has bought for $30 a book from "bookstore.com" on the [EQUATION] of November [EQUATION]. A receipt for this transaction, issued according to our protocol, is Figure 8.2.

Policy Language

Service providers usually evaluate users based on previous transaction interactions with service providers. Based on users' transaction history, service providers are able to determine whether a user can be trusted or whether he can be qualified to gain some benefits such as a discount or rebate. Service providers define policies, referred to as *verification policies*, to specify the conditions against attributes which are recorded in transaction receipts. Verification policies are formally defined as follows.

Term   A *term* is is an expression of the form Name(attribute_list) where Name is the name of a service or discount or an item, whereas attribute_list is a (possibly empty) set of attribute names characterizing the service.

Attribute Condition   An attribute condition, Cond, is an expression of the form: "$name_A$ op l", where $name_A$ is the name of a transaction attribute, A, op is a comparison operator such as =, <, >, Ψ, $\square$, $\neq$, and l is a value that can be assumed by attribute A.

> **Example 4.2** *Some examples of attribute conditions are as follows.*
> - SELLER = *"BookStore.com"*
> - DATE < *"11-04-2008"*
> - PRICE > \$80

Verification Policy   A *verification policy* Pol is an expression of the form

$$\mathcal{R} \leftarrow \text{Cond}_1, \text{Cond}_2, \ldots, \text{Cond}_n, n \geq 1,$$

where $\mathcal{R}$ is a term and Cond_1, Cond_2, … ,Cond_n are attribute conditions.

Given a transaction receipt, Tr, and a verification policy

$$\text{Pol} : \mathcal{R} \leftarrow \text{Cond}_1, \text{Cond}_2, \ldots, \text{Cond}_n, n \geq 1,$$

if for each Cond_i in Pol, $\bar{A} \in \text{Tr}.ATT$ such that $name_{\bar{A}} = \text{Cond}.name$ and $value_{\bar{A}}$ satisfies Cond ($name_A$ op l), we say that Tr satisfies Pol, written as Tr3Pol.

> **Example 4.3** An example of verification policy is the following: Pol: Discount(OnItem ="Glamour", Amount="\$15") U SELLER = "BookStore. com", PRICE > "\$80", DATE < "11-04-2008".
> The policy states that a user is qualified for a \$15 discount on a yearly subscription to *Glamour* magazine, if the user has spent more than \$80 at "BookStore.com" before the date "11-04-2008."

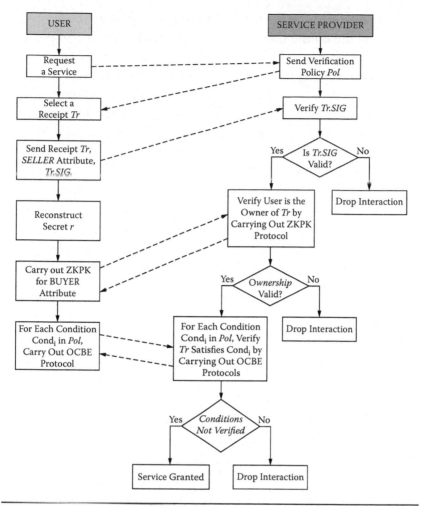

Figure 8.3 Privacy-preserving receipt management.

Privacy-Preserving Receipt Management

Carried out between a user and a service provider, the privacy-preserving protocol proves the possession of a transaction receipt. The protocol consists of four main phases (see Figure 8.3):*

1. **Integrity verification of receipt's attributes.** The user sends a transaction receipt to a service provider to satisfy the service

---

* In what follows we use the term *user*; however, in practice the steps are carried out by the client software transparently to the actual end user.

provider verification policy. The service provider verifies the signature on the transaction receipt sent by the user to prove the satisfiability of service provider's verification policy.

2. **Secret sharing on the mobile device.** The user reconstructs the secret values r that have been used to compute the transaction attribute commitments for the zero-knowledge proof of knowledge protocol, which were split for better protection from unauthorized accesses.

3. **Proof of receipt ownership.** The user proves he or she is the owner of the transaction receipt by carrying out a zero-knowledge proof of knowledge protocol with the service provider.

4. **Verification of conditions on receipts.** The service provider verifies that the transaction receipt attributes satisfy its verification policy by carrying out an OCBE protocol with the user.

In the following sections, we describe in detail each phase of the protocol.

Integrity Verification of Receipt's Attributes This phase starts when a user makes a request to a service provider, and the service provider sends the user the corresponding verification policy $RUCond_1$, $Cond_2$, ... , $Cond_n$, n 1. The user first selects a transaction receipt Tr that satisfies such policy. Then, the user sends the service provider Tr.COM, Tr.SIG, Tr.ATTR.SELLER, the identifier of the service provider which has issued Tr. The service provider retrieves the public key K_{Pub} of the service provider that has issued Tr to be able to verify the signature Tr.SIG.

If Tr.SIG is valid, the service provider verifies that the user is the real owner of the receipt by carrying out a ZKPK protocol with user; otherwise it drops the interaction with the user.

Secret Sharing on the Mobile Phones In order for a user to be able to carry out ZKPK and OCBE protocols with the service provider, the user needs the random secret values r, used to open the Pedersen commitments of a transaction receipt's attributes. The security of the protocols strongly depends on r so it is necessary to protect them from unauthorized access that can occur on mobile devices. Mobile device security can be compromised if the device is lost or stolen, or due to the vulnerabilities of the communication network and/or the device software. To defend these security threats, we adopt Shamir's secret sharing scheme that allows one to split a secret in n shares and then reconstruct it if and only if k shares are present. The storage of the shares depends on the specific architecture of the mobile devices. Next we will focus on the Nokia 6131 NFC mobile phones when describing a design.

In our design, the shares are stored on different mobile phone components and possibly on external devices such as a PC or in an external storage unit. We split each random secret into four shares s_1, s_2, s_3, and s_4. The first share, s_1, is stored in the internal memory of the mobile phone. The second share, s_2, is further split into two secrets. A user chosen PIN number P and a number P' are selected such that $P \frac{1}{2} P' = s_2$, where $\frac{1}{2}$ is bitwise XOR. P' is stored in the phone external memory. The third share, s_3, is stored in the smart card integrated in the phone. Finally, the fourth secret share, s_4, is stored in the user's PC or at an external storage service that has to be accessed remotely by the phone. We consider four levels of protection for the secret, r, that correspond to the number, k, of shares that are needed to reconstruct r. The possible levels of protection are *low*, *medium*, *medium–high*, and *high*. The level of protection, *low*, requires no splitting of the secret, r. In this case, r is stored in the phone smart card. The *medium* level corresponds to a value of k equal to 2. In this case, the user has to retrieve two of the four shares s_1, s_2, s_3, and s_4 to obtain the secret, r. If the *medium–high* level is chosen, three shares are needed, while with a high level of protection, all the four shares are needed to reconstruct the secret.

The level of protection is set by the user once the issuer of a transaction receipt sends the user the random secret r along with the transaction receipt containing the Pedersen commitments computed using r. The specification of the security level and the entering of the PIN are the only steps that need to be carried by the actual end-user. The security level can, however, be set as a default, and the end user does not need to enter it each time it receives a new receipt. Once set, the level of protection should not be changed by the user.

When the user has to prove the ownership of the transaction receipt sent to the service provider, r needs to be reconstructed. In order to do that, a number of shares, according to the level of protection set up by the user, needs to be retrieved and then combined to obtain r.

Example 4.4. Suppose that John Smith has to prove the possession of receipt

```
"“1234",
("John Smith", "BookStore.com", "Books", "$30", "11-04-
2008"),
("BUYER, 45785687994674 ≥, "CATEGORY, 765539408942,
"PRICE, 2223422262≥, "DATE, 583002423412),
13753507485305035637037
```

to service provider "SomeBookStore." In order to accomplish that, John Smith needs to reconstruct the secret r used to compute the Pedersen commitments contained in the receipt. John Smith sets the security level for r to high and to retrieve each secret share he has to perform the following steps:

1. John Smith retrieves s_1 from the phone internal memory.
2. In order to retrieve s_2, John Smith enters the secret PIN number P using the phone's keypad. P' is retrieved from the phone external memory, and it is used to compute the second secret share $s_2 = P\frac{1}{2}P'$.
3. John Smith retrieves the secret s_3 from the phone smart card.
4. To retrieve the secret share s_4, stored at the user's PC, John Smith connects its PC to the phone.

In contrast, if John sets up a medium security level, he has to retrieve only two shares to obtain the secret r. For example, John can decide to get the shares s_1 and s_3 from the phone's internal memory and the phone smart card, respectively, without having to insert any PIN number (see Figure 8.4).

Proof of Receipt Ownership   Once the user has reconstructed the random secret r, the proof of ownership of the transaction receipt can be achieved by engaging a zero-knowledge proof of knowledge protocol in the section titled "Zero-Knowledge Proof of Knowledge" (ZKPK) for the BUYER transaction attribute with the service provider. According to the zero-knowledge proof of knowledge protocol, the user randomly picks y, s in $\{1, \ldots, p\}$, computes $d = g^y h^s$, where g and h are the public parameters of the service provider. The user then sends d to the service provider. Once d is received, the service provider sends back a random challenge $e \in \{1, \ldots, p-1\}$ to the client. Then the user computes $u = y + em$ and $v = s + er$ where m is the value of the BUYER transaction attribute, and r is the random secret, and sends u and v

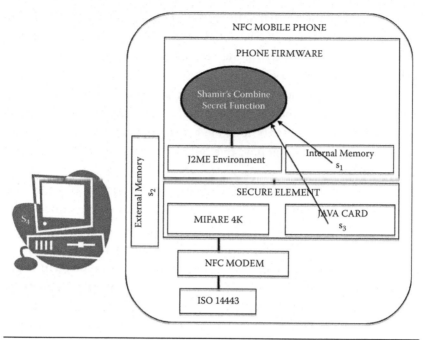

Figure 8.4   Random secret reconstruction.

to the service provider. The service provider accepts the aggregated zero-knowledge proof if $g^u h^v = dc^e$. Otherwise, the interaction with the user is dropped.

Verification of Conditions on Receipts   We consider two scenarios that require the verification of conditions on transaction receipts. In the first scenario, a service provider provides a general service to all qualified users, and does not require knowing the outcome of the transaction. For example, a bookstore may provide a transferable 10%-off coupon code to any user who presents a receipt showing a purchase of a product in the "Books" category. However, the bookstore does not care whether this coupon code is successfully received by the user; it only cares that a coupon code is valid when being used. The book store simply rejects a receipt if it is shown twice to prevent a user from taking advantage of this offer multiple times. In such a scenario, the OCBE protocols (cf. the section titled "OCBE Protocols") can be used directly. Let the user be the receiver Re, and the service provider be the sender Se. Re sends a service request to Se, and Se responds with its verification policy. Based on the policy,

Re selects a receipt Tr that satisfies Se's policy, and sends Tr.COM, Tr.SIG, and the value of SELLER attribute to Se. Se chooses the message M, as described in the section titled "OCBE Protocols," to be the content of service (e.g., a coupon code). Then, it composes the envelope using the corresponding attribute value in the received receipt for M, and sends it to Re. Re can open the envelope if, and only if, the involved attribute value on the receipt satisfies the condition specified in the policy, but Se will not know if Re can open the envelope.

In the second scenario, the service provider needs to know the result of the condition verification; that is, it should be informed if the attributes on the user's receipt satisfies the specified policy. There are many instances of such a scenario. For example, the service provider may require its policy to be satisfied by a user's receipt in order to continue the transactions. In this case, for user privacy protection, the OCBE protocol for equality predicates, EQ-OCBE, should not be employed, because the service provider will be able to infer the attribute value if the verification is successful. However, other OCBE protocols that are for inequality predicates can still be used, with one more step appended to the protocol, described next.

In this additional step, the service provider acts as the sender Se, and the user acts as the receiver Re. The service provider chooses the message M to be a random bit string, which will be used as a secret of Se. The OCBE protocol for inequality predicates is executed between Se and Re, based on Se's policy and the involved attribute value recorded in Re's receipt, for this secret M. At the end of the protocol, after opening the envelope, Re shows Se the decrypted message M'. The attribute on the receipt passes Se's verification if $M = M'$, or fails if it is otherwise. The service provider continues with the transactions in the former case, or aborts the transaction in the latter case. Such an additional step has been added to the OCBE protocols to allow the service provider to learn the result of the verification, at the user's will. Since the random bit string M contains no useful information about the service content itself, a qualified user must choose to show the correctly decrypted secret message M in order to continue the transactions with the service provider. In this sense, the extended OCBE protocols (for inequality predicates) works as a zero-knowledge proof scheme for our application.

In both scenarios, if the user's receipt's attributes need to satisfy multiple conditions in the service provider's policy, a run of the OCBE protocol must be performed for each condition. The procedure, however, can be optimized if the verification is on a conjunction of equality conditions. Interested readers may refer to Reference 17 for details. A receipt's attributes satisfy the conditions in the policy if and only if the user can open all related envelopes.

Implementation Notes

We will describe an implementation of our protocol using Nokia 6131 NFC [12] mobile phones. NFC enabled devices are gaining popularity because they provide easy-to-use mechanisms for ubiquitous accesses to systems and services. Based on a short-range wireless connectivity, the communication is activated by bringing two NFC compatible devices or tags within a few centimeters from one another.

The system architecture of our protocol is shown in Figure 8.5. The core component of this architecture is the Nokia 6131 NFC mobile phone. The Nokia 6131 NFC phone's architecture has been reviewed in the section titled "Nokia 6131 NFC Phone." It consists of an Antenna, an NFC modem, a Secure element, phone firmware, a SIM card, a J2ME environment included in phone firmware, and External memory.

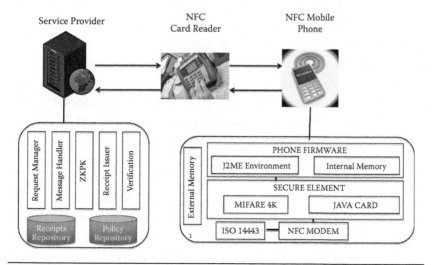

Figure 8.5 System architecture.

The NFC reader enables the communication between the service provider application and the mobile phone. It transmits and receives messages from the NFC cellular phone. The service provider application consists of five main modules: Request Manager, Message Handler, ZKPK (zero-knowledge proof of knowledge), Receipt Issuance, and Verification. The Request Manager module parses user requests and selects from a local repository the verification policy that applies to the request. The Message Handler module provides all functions supporting the communications between the service provider application and the external NFC reader. The ZKPK module supports the verification of receipts' integrity and the ZKPK protocol to verify the BUYER attribute. The Receipt Issuance module provides the functions for creating a transaction receipt, such as the generation of the Pedersen commitments and the signature of the commitments. Once created, the transaction receipts are stored in a local repository. The Verification module supports the steps for the OCBE verification of conditions on receipts described in the section titled "Verifications of Conditions on Receipts."

The transaction receipts will be stored in the external phone memory, whereas the secret r used to compute the secure commitments included in the receipts are saved in the Java Card component. The execution of the MIDlet is triggered when the Mifare 4K captures the verification policy sent by the service provider's external NFC reader and the Mifare 4K transfers such policy to the phone main memory. The MIDlet retrieves from the external memory a transaction receipt that satisfies the service provider policy and sends the part of the receipt containing the transaction attributes commitments, the signature affixed on the commitments, and the value of SELLER attribute to Mifare 4K so that can be read by the service provider's external NFC reader. If the service provider application successfully verifies the signature on the receipts commitments, the MIDlet retrieves the secret r from the Java Card, and performs the other steps of the receipts management protocol.

The MIDlet runs on Java 2 Micro Edition (J2ME). Since J2ME is aimed at hardware with limited resources, it contains a minimum set of class libraries for specific types of hardware. For implementation on conventional nonmobile platforms, the java.math.BigInteger and java.security.SecureRandom classes can be used to implement secure commitments. But both packages are not supported in J2ME; for this purpose, the third-party cryptography provider Bouncy Castle [8], a lightweight cryptography APIs for Java and C# that provide implementation of the BigInteger and SecureRandom classes. In addition, because of the limited memory size of mobile phones, it is helpful to reduce the MIDlet's code size by using code obfuscation techniques provided by Sun's NetBeans IDE. Code obfuscation allows one to reduce a file size by replacing all Java packages and class names with meaningless characters. For example, a file of a size of 844KB can be reduced to a size of 17KB.

Integrity Verification of Receipt Attributes

The integrity verification of receipt attributes step (section titled "Integrity Verification of Receipts Attributes") can be implemented with any secure digital signature algorithms, such as RSA [7], DSA, and ECDSA [5]. Implementation of many digital signature algorithms are readily available in the Bouncy Castle cryptography API, which can be conveniently used in Nokia 6131 NFC phone's J2ME environment. Examples are the org.bouncycastle.crypto. signers.RSADigestSigner and org.bouncycastle.crypto.signers. DSADigestSigner classes. If RSA signatures are chosen to implement the receipt attribute integrity verification, we recommend to set the RSA modulus size to 2048-bit or higher and use SHA-256 or stronger as hash digest.

Secret Sharing on the Mobile Phones

The secret sharing implementation should be compatible with the Pedersen commitment parameters. In particular, the size of the finite field F_p over which the polynomials are defined should be large enough so that the secret exponent r in any Pedersen commitment can be lossless-encoded as a field element. When 112-bit key strength is satisfied, r will be around 224 bits. In this case, the finite field used for the secret sharing algorithm should be larger than 224 bits. Choosing p to be a prime (so that the finite field is a prime field) perhaps is most convenient for the implementation, but this is not absolutely necessary. Also, the PIN number used to reconstruct the secret should be long and random enough for strong protection.

Proof of Receipt Ownership

The privacy-preserving ownership proof of the receipts are performed between the service provider and the user (mobile device) via the zero-knowledge proof of knowledge protocol (section titled "Proof of Receipt Ownership"). A key component of the proof, the Pedersen commitment scheme, is described in the section titled "Pedersen Commitment" in terms of a generic cyclic group. Suitable instantiations of such a cyclic group are subgroups of the multiplicative group of a finite field, or subgroups of elliptic curves over finite fields. With the Bouncy Castle library for the J2ME environment, it probably is more straightforward to implement the former, using the org.bouncycastle.util.BigIntegers class with modular arithmetic. In this case, the parameters can be chosen as in DSA. We recommend to choose g and h as elements of 2048 bits, whose order modulo a 2048-bit prime is a 224-bit prime p, to meet the 112-bit key strength requirement.

An illustrative implementation of the zero-knowledge proof of knowledge protocol (with [EQUATION]-bit key strength) shows that it takes on average 42 and 31 milliseconds for the proof, on the mobile phone's MIDlet and the service provider's desktop application, respectively.

Verification of Conditions on Receipts

As in the case of zero-knowledge proof of knowledge, the OCBE protocols for privacy-preserving verification of conditions also have the Pedersen commitment scheme as a crucial building block. The implementation is similar in both cases. Note that for security, the parameter sets $"G, g, h$ should be distinct for different attributes, and should be different from the one used for ownership verifications.

An experiment shows in Tables 8.1 and 8.2 the average running time of OCBE protocols on the mobile phone's MIDlet and the

Table 8.1   Verification of condition execution time (in seconds) at MIDlet's side (, =5)

	COMMITMENTS CREATION	OPENING ENVELOPE	TOTAL EXECUTION TIME
Equality condition	0	1.126	1.126
Inequality condition ( )	5.875	6.088	11.963

Table 8.2 Service provider's application's average execution time (in seconds) for verifying one condition (, = 5)

	ENVELOPE CREATION
Equality condition	0.0409
Inequality condition (≥)	0.165

service provider's desktop application, respectively. The parameter , = 5 for data presented in both tables. More comprehensive comparisons are shown in Figures 8.6 and 8.7. For efficient implementation of our protocol, the parameter , must be kept as small as possible in order to reduce the computational cost, while at the same time it should be large enough to accommodate all involved attribute values.

Development with Nokia 6131 NFC SDK

The Nokia 6131 NFC SDK [11], along with other things, implements the *Contactless Communication* API *(JSR-257)* [6], which enables the NFC features of the phone. The P2PExample MIDlet found in the Nokia 6131 NFC SDK demonstrates how to communicate between two Nokia 6131 NFC devices using the com.nokia.nfc.p2p.NFCIPConnection interface. The communication part of the privacy-preserving receipt management protocols (attribute verification, ZKPK and OCBE) between the service provider and

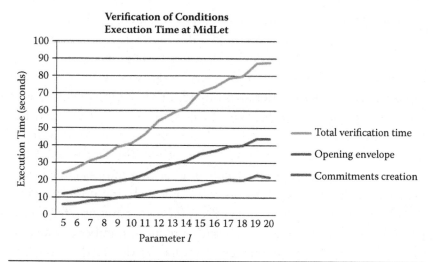

Figure 8.6 MIDlet's Envelope Opening Time, varying the value of parameter ,.

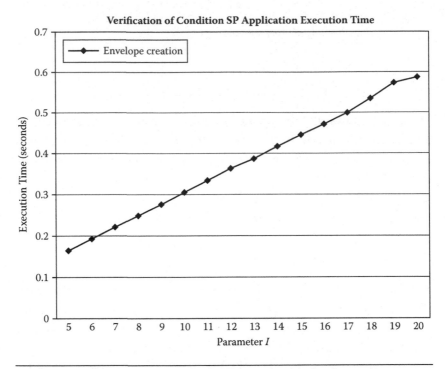

Figure 8.7 SP Application's Envelope Creation Time, varying the value of parameter ,.

the user can be implemented in a way similar to the SDK example. In the peer-to-peer mode, one device should be set as *initiator* and the other as *target*, which can be done by opening the connection (by calling javax.microedition.io.Connector.open) with strings "nf c:rf;type=nfcip;mode=initiator" and "nfc:rf;type=nfcip;mode=target," respectively. The *initiator* device first sends data and then receives a response from the *target* device; the *target* device does it in the opposite way. For all aforementioned privacy-preserving receipt management protocols, the user device will be an *initiator* and the service provider will be a *target*. Note that a connection is initiated when a device is put close to the other. The fact that ZKPK and OCBE are interactive protocols implies that multiple touches or taps are needed for the protocols to complete. Using noninteractive zero-knowledge proof techniques [2,4] for ownership proof makes it possible to achieve the same goal with a reduced number of touches, but will also make the implementation harder due to the increased complexity of the protocols.

It is challenging to make the peer-to-peer NFC connection work between an NFC phone and an NFC reader due to the lack of compatible NFC readers that supports this mode. A workaround is to connect the second NFC phone with the service provider's computer (through Wi-Fi, Bluetooth or wire) and to let the service provider use this NFC phone as a proxy to communicate with the user's NFC device.

Conclusions

We have proposed a privacy-preserving method and a software design to manage electronic transaction receipts stored on NFC mobile devices. Our approach is based on the notion of *transaction receipts*, and combines various cryptographic techniques to achieve the goal of user privacy preservation. Our design is specific for the Nokia 6131 NFC mobile phone. We also give suggestions and recommendations regarding the implementation of the system. We believe mobile devices equipped with communication technologies such as NFC will have a bright future.

References

1. Help for lost and stolen phones. http://news.bbc.co.uk/1/hi/technology/4033461.stm.
2. M. Belenkiy, M. Chase, M. Kohlweiss, and A. Lysyanskaya. P-signatures and noninteractive anonymous credentials. In *TCC*, pages 356–374, 2008.
3. W. Gautschi. *Numerical Analysis: An Introduction*. Birkhauser Boston Inc., Cambridge, MA, 1997.
4. J. Groth and A. Sahai. Efficient non-interactive proof systems for bilinear groups. In *EUROCRYPT*, pages 415–432, 2008.
5. National Institute of Standards Information Technology Laboratory and Technology. FIPS PUB 186-3: Digital Signature Standard (DSS), 2009.
6. Java specification requests: Contactless communication API. http://jcp.org/en/jsr/detail?id=257.
7. RSA Laboratories. PKCS#1 v2.1: RSA Cryptography Standard, 2002.
8. Bouncy Castle Crypto APIs. http://www.bouncycastle.org/.
9. J. Li and N. Li. OACerts: Oblivious attribute certificates. *IEEE Transactions on Dependable and Secure Computing*, 3(4):340–352, 2006.
10. Near Field Communication Forum. http://www.nfc-forum.org.
11. Nokia 6131 NFC SDK 1.1. http://www.forum.nokia.com/info/sw.nokia.com/id/ef4e1bc9-d220-400c-a41d-b3d56349e984/Nokia_6131_NFC_SDK.html, 2007.
12. Nokia 6131 NFC. http://wiki.forum.nokia.com/index.php/Nokia_6131_NFC_-_FAQs.

13. F. Paci, N. Shang, S. Kerr, K. Steuer Jr., J. Woo, and E. Bertino. Privacy-preserving management of transactions' receipts for mobile environments. In *IDtrust*, pages 73–84, 2009.
14. T.P. Pedersen. Non-interactive and information-theoretic secure verifiable secret sharing. In *CRYPTO '91: Proceedings of the 11th Annual International Cryptology Conference on Advances in Cryptology*, pages 129–140, London, 1992. Springer-Verlag.
15. C-P Schnorr. Efficient identification and signatures for smart cards. In *CRYPTO '89: Proceedings of the 9th Annual International Cryptology Conference on Advances in Cryptology*, pages 239–252, London, 1990. Springer-Verlag.
16. A. Shamir. How to share a secret. *Commun. ACM*, 22(11):612–613, 1979.
17. N. Shang, F. Paci, and E. Bertino. Efficient and privacy-preserving enforcement of attribute-based access control. In *IDtrust*, pages 63–68, 2010.
18. TechRepublic. Identify and reduce mobile device security risks. http://articles.techrepublic.com/5100-22_11-5274902.html

9

PERFORMANCE OF AN NFC-BASED MOBILE AUTHENTICATION AND PAYMENT SERVICE CONCEPT COMPARED WITH TRADITIONAL SOLUTIONS

MICHAEL MASSOTH AND THOMAS BINGEL

Contents

This contribution presents a performance comparison of four traditional mobile payment service concepts with a state-of-the-art solution based on Near Field Communication (NFC). The focus lies on the concept itself rather than the implemented software. The NFC application is compared in a benchmark with Interactive Voice Response (IVR), Short Message Service (SMS), Wireless Application Protocol 2.0 (WAP), and a one-time password (OTP) generator.

Introduction

An increasing number of mobile devices are produced with modern technologies built in. Especially, the so-called smart phones are equipped with a lot more than only a Global System for Mobile Communications (GSM) and General Packet Radio Service (GPRS) module. In our opinion, the technology of NFC (Near Field Communication) is especially very attractive.

Customers could use these new technologies for their mobile payment, although the performance of mobile payment methods is an essential factor to be considered. Therefore, we present a performance measurement of five mobile payment applications. These benchmarks are shown in an end-to-end service duration time point of view.

The GSM and 3rd Generation Partnership Project (3GPP) already have embedded mature authentication protocols. However, the customer usually has to transfer an additional personal identification number (PIN) or password to authorize a mobile commerce transaction in mobile payment.

Authentication

Authentication, in the sense of computer science, describes the process to obtain, verify, and prove the identity of a user or service. The goal is to assure that you are really dealing with the alleged user or service. The most common method of authenticating a user is to request the password. However, this is only one example of all possible kinds of authentication methods. They could be categorized into four groups as follows:

- *Something you know*: The customer has to authenticate himself with some information he and the authenticating party share.

The most common type of such information is a password. Other examples are a pass phrase or PIN.

- *Something you have*: The customer has to show a specific thing i.e., a card, tag, or token, he possesses during this mechanism. This type of authentication is common in the financial sector, where the customers are identified by their credit cards. It is important that the authenticating party be able to prove that this thing is the real one.

- *Something you are*: The property or knowledge of a person is a little easier in comparison to stealing or copying. The authentication could be coupled to a unique feature of the customer himself to reduce this risk. This could be a fingerprint, retina or vein pattern for example. The disadvantage is if a criminal is still able to copy this feature, that it could not be changed any more.

- *Somebody you know*: It is possible to adopt a new form of authentication with a growing distribution of social networks. It enables authentication of a user or service if another vouches for him or it [1]. This concept is also known as Identity 2.0 and will be used in social networks.

Authentication is much better if these mechanisms are performed in combination. Unfortunately, such a combination is more complicated to move ahead. This leads to a longer duration for an authentication, which is only partially possible for mobile payment.

Let us give an example for an authentication process. The best-known payment example is the use of a credit card. A customer who wants to buy a new product in a retail store and does not have enough cash could pay by using his credit card. Therefore, the cashier wants him to enter his PIN and sign the bill. Then this signature will be compared with that on the credit card. This process includes three forms of authentication mechanisms:

- Something you have: The credit card.
- Something you know: The PIN.
- Something you are: The signature.

This process takes some time. Therefore, most cashiers do not want all three authentication mechanisms. It is usual to only sign the bill and compare the signature with the one on the card. However, the risk of fraud is much higher if the PIN is not entered.

The example above also illustrates the threat of an insecure authentication. Assume the customer is not the real owner of the credit card. It is rather simple to copy the signature from the credit card. This customer is able to spend all the money from the account without the authentication via PIN until the card is disabled by the financial institution [2,4,5].

Proximity Authentication

The example just presented describes proximity authentication. It describes the distance between both participating parties during the authentication process. The customer gives a credit card to the cashier and he swipes it through the reader. The cashier could check whether the credit card is valid and not some kind of fake. Also, he could compare the signatures. Proximity authentication is already used by many traditional payment systems such as credit cards, bank billing, or bank collection.

Remote Authentication

Another type of authentication is *remote authentication,* where someone sits at home at a computer or another kind of terminal and has to authenticate himself. The participant is not in close range to a person who could check if the authentication is valid. This makes the authentication very difficult. Especially for the authentication mechanism *"Something you are."* Who has a fingerprint or retina scanner at home right next to the computer? The basic remote authentication mechanism is *"something you know,"* which most of the time is a password containing a number-character combination of at least eight random characters. A short password or a password containing a well-known word is called a *weak password* because it can be guessed very easily. *"Something you have"* is harder to cover with a remote authentication such as Internet payment compared to proximity authentication. A criminal could steal a credit card and use it for Internet shopping. No one will ask him for the PIN or signature. The number printed on the card is enough to authenticate it. But because he has a credit card does not mean that he is its owner! And even not having the physical card is important with these online transactions; only the numbers printed on the front and the back of the card!

Industrial Authentication Mechanisms

As a result of the vulnerability of traditional payment systems, new ones had to be provided to pay online. Paypal, Google Checkout, Bill Me Later, and Money Card are only some of the bigger companies providing such services. All these services use the Internet to enable customers to make payments. These payments are almost invariably Internet-based and not payments such as in retail stores.

For an online payment the consumer has to enter a user ID and a password to log into his account. Then he or she can transfer money to another registered consumer. The big problem with all the Internet payments is that the authentication mechanism is still *"Something you know"* and not *"Something you have"* nor *"Something you are."*

The computer could be compromised from the beginning of a payment and spyware could log every key being pressed. This would mean that someone knows your user ID and password. He could transfer money to another account and then to many other accounts around the world making it nearly impossible to trace.

Challenges for an Authentication

The computer cannot be seen as trustworthy. Trojans, spyware, viruses, or other kinds of malicious software (*"malware"*) can compromise computers. This is where the paybox comes into play, where the mobile phone is a trusted device representing *"Something you have."* Also, the PIN which has to be entered to authorize a payment, representing *"Something you know"* makes this payment method very secure.

There are many threats making authentication difficult, with *"something you have"* being stolen, *"something you know"* being eavesdropped, or *"something you are"* being imitated. Especially on the computer there are many threats possible, such as man-in-the-middle attacks, phishing, or replay attacks. It is the task of the payment provider to prevent such attacks by implementing security means. Still, the biggest security risk exists right in front of the monitor, operating the computer. Therefore, it is a good idea to migrate payment to another device such as a mobile phone, where the user does not have many options to break something or to install spyware by mistake.

But this could change in the future if computers and mobile devices fuse even more and spyware for mobile devices becomes more widespread. However, as Tomi Dahlberg et al. pointed out, *"The impacts of social and cultural factors on mobile payments, as well as comparisons between mobile and traditional payment services are entirely uninvestigated issues"* [3].

Authorization

Authorization defines which person or program is permitted to have access to specific data or functionalities. Sometimes, an authentication is additionally performed during this process. Users usually do not have a lot of privileges if they are not authenticated, because they are restricted to limited or no resources. Authorization allows another party to do something such as transferring money from one account to another, or allowing access to a restricted area. It protects resources but only allows those resources to be used by someone that has the granted authority to use them. The better the authentication is, the more valuable the resources. The following mobile payment concepts try to combine proximity and remote authentication to authorize a payment by using a new medium: the mobile phone as a payment device or mobile wallet.

The Mobile Authentication and Payment Concepts

All mobile authentication and payment concepts have been developed and realized together with the former Paybox Solutions AG in Germany; not just the demo applications were developed.

- The paybox to consumer gateway (e-mail and SMS [Short Message Service] gateway) was enhanced with the possibility of sending not only SMS messages, but also configuring them with a push port. This enables the paybox to start Java Platform, Micro Edition (JAVA ME) applications on the mobile phone.
- Also, WAP Push was implemented for the gateway to enable an easy way to deliver links to the JAVA ME application.
- Another application was developed to send all these different message types not only over an SMS gateway but also over a regular mobile phone.

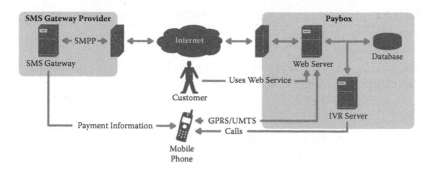

Figure 9.1   Overview of the mobile payment test system.

- Each demo application is listed in a newly created paybox research lab web page. The goal of this web page is to create a *"mobile payment playground"* with all the different application and show how they work.
- Thus, merchants can look at the different options and can decide what kind of payment service they would like to use.

Figure 9.1 shows the complete architecture of the demo system. It consists of a web server, an IVR (Interactive voice response) server, and a database. The SMS gateway is provided by another company allowing access over Short Message Peer to Peer (SMPP).

Table 9.1 lists all demo applications with the technologies and authentication methods used.

As shown, all mobile payment application concepts, except the SMS application, have two authentication methods, and therefore provide strong security.

Table 9.1   Overview of all mobile payment application concepts

APPLICATION	TECHNOLOGIES	AUTHENTICATION METHOD
IVR	IVR (Envox), Java Platform, Enterprise Edition (JAVA EE), Oracle	*"Something you know"* *"Something you have"*
SMS	SMS gateway, JAVA EE, Oracle	*"Something you have"*
OTP generator	SMS gateway (WAP Push), JAVA EE, JAVA ME, GPRS, MySQL	*"Something you know"* *"Something you have"*
NFC	NFC, JAVA EE, JAVA ME, GPRS, MySQL	*"Something you know"* *"Something you have"*
WAP	SMS gateway (WAP Push), JAVA EE, GPRS (WAP), MySQL	*"Something you know"* *"Something you have"*

Table 9.2   Use case for general registration to all mobile payment application concepts

Name	Registration	
Initiator	Consumer	
Goal	Register the mobile phone in the paybox demo system.	
Participants	Consumer, paybox	
STEP	ACTOR	DESCRIPTION
1	Consumer	Wants to register at the paybox demo system.
2	Consumer	Visits the paybox register website.
3	Consumer	Enters phone number, PIN, and phone type.
4	Consumer	Submits data.
5	Paybox	Confirms registration and saves data into the database.
6	Paybox	The consumer is now registered and ready to access the payment demo applications.
7	Paybox	Sends SMS to the consumer as a confirmation.
Registration successfully completed		

General Registration

All demo applications need the same basic information to be able to authenticate a consumer at the payment system. These are the following:

1. The mobile phone number, which basically is the identification of the consumer.
2. The PIN, which authenticates the consumer at the system.
3. Information about the used mobile phone. This helps to create customized applications for the mobile phone.

JAVA ME applications, especially, rely on customization because they behave very differently on various phones. Bank information is not required for the demo applications. Table 9.2 contains the use case describing the registration process.

Interactive Voice Response

IVR stands for an automated telephony system to communicate with a person on the other end of the telephone line. Usually, the IVR system plays back prerecorded voice files to give the caller instructions what to do next and how to interact with the system. The IVR authentication

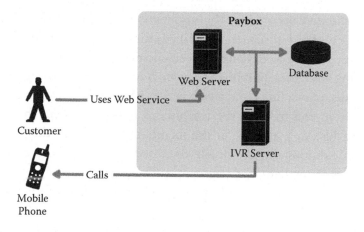

Figure 9.2 Components of the IVR mobile payment system.

application uses the basic GSM mobile telephone technology to call a consumer and ask for his PIN (see Figure 9.2 and Figure 9.3).

When buying a product, the customer selects the payment option "IVR" and enters his mobile phone number. Then the IVR authentication application initializes an IVR call to the customer and asks him or her to enter the PIN. The consumer enters the PIN, and the payment is completed.

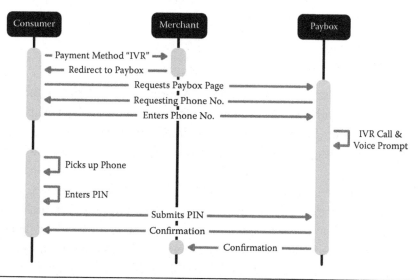

Figure 9.3 Sequence diagram of the IVR mobile payment system.

The components used for this application include the following.

An IVR server is making the calls to the consumer. The server currently uses Envox to develop the voice logic. The server also includes recording voice files, building the voice flow, specific hardware for the voice calls, phone lines, and a web server (Apache Tomcat) to provide a web front-end to enter the consumer's phone number. The server also provides web services for the merchant to initialize the payment process. Additionally, an Oracle database is used to store customer data, transaction data, and payment information.

The time to process a call depends on how much information or choices the IVR system gives to the caller and how fast the caller answers the questions. A single channel will be busy for that amount of time and cannot be used for another caller. The more time a caller takes to answer the question, the more expensive will be the phone bill at the end and the fewer the callers that can be processed. The IVR solution is able to provide remote and proximity payments. The consumer must have network reception. This can be a problem in retail stores, where sometimes there is no reception at all. Also, the communication between cashier and consumer to exchange the phone number can be tricky. Some people do not know their phone number by heart. The best way to use an IVR system is at home, where a consumer sits in front of the computer and wants to pay for a product online (remote payment).

Short Message Service

The SMS provides a way to send Short Messages (SMs) between mobile devices. To send SMs from one device to another, a Short Message Service Center (SMSC) is needed. It is responsible for receiving and delivering messages by means of a store and forwarding mechanism. If a device sends an SM to another device, it connects to the SMSC, which stores the SM and tries to deliver the SM to the receiver. The SMSC will try to deliver the message for only a certain time, which can be configured on the sender side up to one week. If it cannot be delivered in that time, it will be deleted. There is no direct communication between sender and receiver. It is possible to get a confirmation message that the SM was delivered for an extra fee. In general, all SMs are delivered within a couple of seconds worldwide, as long the receiver's

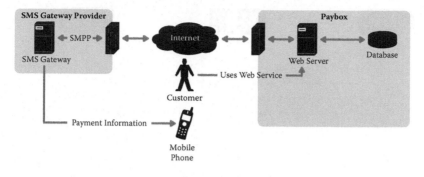

Figure 9.4   Components of the SMS mobile payment system.

device is online. The Push SM is another type of SM, which is able to start a preinstalled application on a mobile device. This is done by registering the application in a registry, called Push-Registry, which was introduced in the MIDlet 2.0 specification for Java ME. The registration is a number-value pair, which specifies the port on which the SM has to be sent and the program that should be started. It is only possible to specify the port when the application is being installed or when the application is started. It is also important to choose a port that is not already used by another application or service such as WAP Push.

The installed application does not run in the background all the time because this would waste a lot of resources. The device checks the Push-Registry every time a SM is received. If the SM is directed to a registered port, it will trigger the application to start.

When the application starts, it is possible to process the content of the SM that triggered the application start. Also, other incoming messages directed to the same port can be processed within the application while running.

The SMS authentication application sends a short message to the consumer asking for an authentication by replying to the message with the content *"YES"* (see Figures 9.4 and 9.5).

The SMS application can be used in remote and in proximity situations. This payment method has the same issues as the IVR method because the payer has to have network reception. This can be a problem in closed rooms such as retail stores. A real-life scenario where an SMS payment is possible is paying for ring tones or phone games where the prize is very small and risk of fraud is very little.

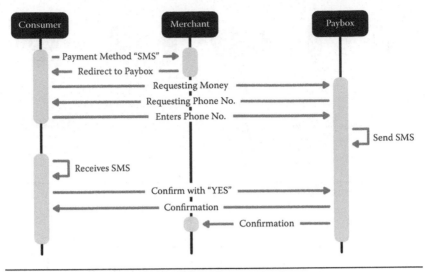

Figure 9.5 Sequence diagram of the SMS mobile payment system.

Wireless Application Protocol 2.0

- The Wireless Application Protocol 2.0 (WAP 2.0) enables the mobile phone to access Internet content. The WAP authentication application sends a WAP Push message containing a customized Uniform Resource Identifier (URI) to the consumer. The consumer opens the message, which starts the WAP browser loading a web page asking for authentication with the PIN (see Figure 9.6 and Figure 9.7).

Components for the WAP application are:

- Web server (Apache Tomcat), which provides a web interface for the consumer to enter the phone number. It also makes

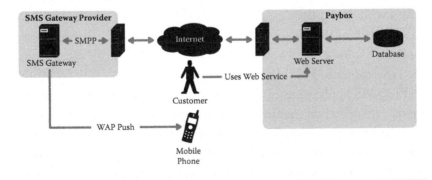

Figure 9.6 Components of the WAP mobile payment system.

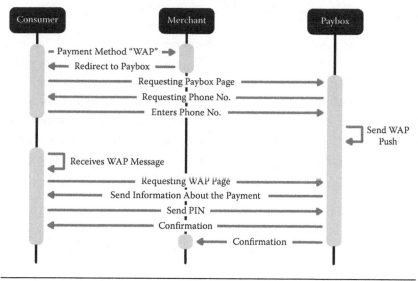

Figure 9.7   Sequence diagram of the WAP mobile payment system.

it possible to process the WAP requests coming from the mobile phone.

- SMS gateway to send a WAP Push message containing the URL to the WAP page to the consumer.

A consumer wants to buy a product via Internet. He or she selects the payment method "WAP." The merchant redirects the consumer to the paybox WAP page, where the consumer has to enter the phone number (MSISDN). Paybox sends a WAP Push message to the entered number, which includes the URL to this specific payment. The SM takes about 10 s to reach the receiver. The consumer accepts the WAP Push message, which opens up the WAP browser on the phone and opens the URL stored in the WAP message. A WAP page opens, displaying a personal greeting (preventing phishing attacks) and the payment information. The consumer can then enter the PIN in an input field for verification. After submitting the PIN, the consumer gets a confirmation message telling him that the payment was successful. The operational area for the WAP application is remote; the consumer needs network reception to use the service.

One-Time Password Generator

A one-time password (OTP) is a secret phrase that is generated for only one usage. The tested OTP generator generates this password by using time synchronization between a paybox server and the used mobile phone. A customer is identified from the contacted server by a hash build from the PIN of the customer and additional information. The customer has to enter this PIN after launching the OTP application. The OTP generator has four main components:

1. The OTP generator itself. It is a JAVA ME application that runs on a mobile device and generates passwords.
2. A file server that distributes user specific JAVA ME applications.
3. A Java synchronization server that enables the mobile device to synchronize the OTP application with the paybox server.
4. A Java-based authentication server that validates the transmitted phone number and OTP.

The requirement to use the OTP authentication method is that the consumer be registered at the paybox system with his phone number and corresponding PIN. He can download an OTP application for his mobile device after registration. The application is delivered by WAP Push, which was also developed during this contribution to provide easy access to the OTP application.

WAP Push messages are specially formatted SMS messages that show up on the display of the mobile device. It gives a customer the option to directly connect the corresponding paybox server using a URI stored in the WAP Push message. The benefit of WAP Push messages is that it is not necessary to enter the inconvenient URI into a mobile device.

A consumer requests the OTP delivery web page, then enters and sends his phone number. The paybox system checks if the requesting consumer is already registered and sends a WAP Push message to the transmitted phone number. This message is received at the mobile device after some seconds and contains a URI, which is easy to use by pressing only a couple of buttons (usually one or two). The URI stored in the WAP Push message identifies a dynamically built Java

Application Descriptor (JAD) file, which adds the following user specific configurations:

1. The phone number, which is part of the encrypted OTP. Unfortunately, this value could not be retrieved by the JAVA ME framework because a lot of mobile devices do not support it.
2. A URI that identifies a synchronization server.

With personalized configurations, it is possible to download an application and start it without doing any manual configuration on a mobile device. When the OTP generator is started, it requests the user to enter his PIN. It is part of the encrypted OTP that will be displayed as soon as the consumer confirms his PIN.

The OTP generator application provides two authentication mechanisms. First, "Something you know," which is the PIN the consumer has to enter into the Java ME application. Second, "Something you have," which is the personalized OTP generator on the mobile device.

The OTP generator can be used in remote as well as in near-by areas. It was designed to be used for an Internet shop where the consumer can enter the OTP on a web page. Also, it can be used on a vending machine containing a modem and a numeric keypad.

One main benefit of the OTP is that the consumer does not have to have network reception. The OTP generator runs without communicating with the server, as long as it was synchronized once.

Near Field Communication

The NFC module of a mobile device reads a point of sales (POS) identity from radio-frequency identification (RFID) tags. NFC is build into more and more mobile devices and becomes popular even for public transportation. A device starts the NFC application after it reads the RFID tag. This application tries to connect to a paybox server and retrieves authentication data. A consumer has to enter his PIN to authenticate himself to this paybox server. The following components (Figure 9.8) are involved:

• A web server that is implemented with Apache Tomcat. It has to offer a web service for a merchant to create open transactions and process payment requests from the NFC application.

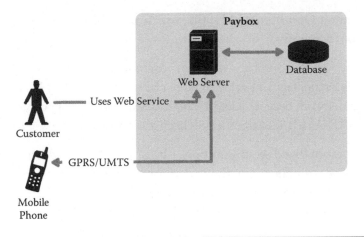

Figure 9.8   Components of the NFC mobile payment system.

- The NFC application itself, which was realized in JAVA ME. It runs on the mobile device, has to read RFID tags, and contact the web server.
- The register that communicates with the paybox server creating transactions.
- A RFID tag that is on the register and contains the identity of the register.

The procedure if a customer buys a product is depicted in Figure 9.9. The cashier creates an open transaction at a paybox server after this customer tells him that he wants to pay via NFC.

After that, the consumer touches an RFID tag at the cashier and the NFC application starts, connecting to the paybox and asking for open payments for this cashier. If the data is correct, the consumer enters the PIN on the phone, which is checked by the paybox. If everything is correct, the cashier gets a confirmation.

The operational area for the NFC application is only a proximity payment such as a retail store. The consumer needs network reception to use this payment method. The communication with the server uses a Hypertext Transfer Protocol Secure (HTTPS) connection to the web server. The transmitted data is a combination of the consumer's phone number, the PIN, and the transaction identifier. All data is hashed with a Message-Digest algorithm 5 (MD5). The transaction identity within the data package creates a unique hash, preventing

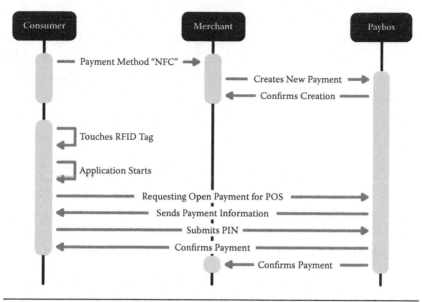

Figure 9.9   Sequence diagram of the NFC mobile payment system.

replay attacks. The authentication server uses the same data to generate a hash value and compares both values. The PIN is encrypted twice on the mobile device because the PIN is stored in the database as an MD5 hash.

Performance Tests

System under Test

The system under test where all demo mobile payment application concepts run is depicted in Figure 9.1. A web server accepts all requests made by the consumer, who can choose which application to test. As soon as the consumer has chosen an authentication method, a demo shop is displayed, asking for the required information.

The database is the central storage for all information, including the user information as well as the data for each transaction. Next is the SMS gateway, which is used in almost every demo application. Last is the IVR server, which is used for the IVR application.

Testing will show how long it takes to make a successful authentication and mobile payment with each application concept. All test cases start with choosing a payment method. The mobile phone has to be in idle mode, meaning no applications are running on the phone. This assures that each test starts from the beginning and the time to start up an application is included. The test data for IVR and SMS are provided by the currently running authentication system of the Paybox Solutions AG.

Evaluation

Table 9.3 shows the end-to-end service time benchmarking results of each mobile authentication application. The speed of an authentication application is hereby defined by the time a single transaction takes and by how many transactions can be done within a certain time span. This can be measured with different statistical criteria.

- NFC is one of the fastest applications because the consumer has only to touch an RFID tag to automatically start the authentication application on the phone. It is not necessary to wait for an SMS or WAP Push message to start the application. The variance shown in Table 9.3 is very low compared to IVR or SMS. This is because the user has to do very little manual work such as starting the application, etc. Still, it can take a couple of seconds to establish the connection to the Internet and to the server.
- OTP is the second-fastest application, because it does not need an Internet connection established at the mobile phone. This saves valuable time and cost. It still takes a long time to start up the application. The high variance comes from the

Table 9.3   End-to-end service time measurements for the different demo mobile payment application concepts

	IVR	SMS	WAP	NFC	OTP
Average	35.60 s	27.93 s	38.14 s	21.98 s	26.43 s
Median	36.00 s	28.00 s	39.00 s	22.00 s	26.00 s
Minimum	25.00 s	20.00 s	29.00 s	18.00 s	18.00 s
Maximum	39.00 s	36.00 s	47.00 s	26.00 s	35.00 s
80% percentile	38.00 s	34.00 s	44.00 s	24.00 s	29.00 s
90% percentile	39.00 s	35.00 s	45.00 s	25.00 s	31.00 s
95% percentile	39.00 s	35.10 s	46.10 s	26.00 s	32.00 s
Standard deviation	2.98 s	5.08 s	5.51 s	2.39 s	3.29 s
Variance	8.86 s^2	25.80 s^2	30.37 s^2	5.73 s^2	10.80 s^2

manual work the consumer has to do: starting up the application and entering the PIN and OTP. Also, if the OTP is about to expire, the consumer has to wait a moment for a new OTP to be generated.

- The third place (for speed) goes to the SMS application. It takes about 6 to 10 s to send the SM to the consumer, which has to be answered with "yes." This takes about 5 to 10 s depending on the consumer. The response takes another 6 to 10 s. The variance is very high compared to the other applications; this is because consumers have different skills in using the SMS features of the phone. Also, the time to send an SMS message varies.
- In the fourth place is IVR. Ten seconds are needed to call the consumer. The authentication announcement takes about 22 s depending on the consumer's name and the authentication information. Last, the consumer has to enter the PIN, which takes about 3 to 5 s. The whole process of authentication takes about the same time for each authentication.
- The last place is WAP. Sending the WAP Push message to the mobile device takes about 6 to 10 s. Opening up the WAP browser and connecting to the Internet takes about 15 to 20 s, entering the PIN another 5 s. WAP has the highest variance of all. This is because there are many factors that determine how fast the application runs, such as the WAP Push message, opening the WAP browser, connecting to the Internet, the connection speed, and entering the PIN.

Conclusion

The performance of different traditional mobile payment service concepts has been compared with a state-of-the-art NFC-based mobile payment solution. Lesson learned:

- NFC is, with regard to speed, security, and usability, a good choice for mobile payment.
- NFC is going to be one of the major technologies in mobile payment and maybe the most likely candidate for a mobile

commerce killer application in the near future as soon as more phones have this technology built in.

References

1. John Brainard, Ari Juels, Ronald L. Rivest, Michael Szydlo, and Moti Yung. "Fourth-factor authentication: somebody you know." *The 13th ACM Conference on Computer and Communications Security.* New York: ACM, 2006. 168–178.
2. Karl Czajkowski, Carl Kesselman, Steven Fitzgerald, and Ian Foster. "Grid information services for distributed resource sharing." *The 10th IEEE International Symposium on High Performance Distributed Computing.* Washington, DC: IEEE Computer Society, 2001. 181–194.
3. Tomi Dahlberg, Niina Mallat, Jan Ondrus, und Agnieszka Zmijewska. "Past, present and future of mobile payments research: A literature review." *Electronic Commerce Research and Applications*, Nr. 7(2) (2008): 165–181.
4. Ian Foster and Carl Kesselman. *The Grid: Blueprint for a New Computing.* San Francisco: Morgan Kaufmann, 1999.
5. Ian Foster, Carl Kesselman, Jeffrey M. Nick, and Steven Tuecke. The physiology of the grid, in *Grid Computing: Making the Global Infrastructure a Reality*, edited by F. Berman, G. Fox, and T. Hey, Chichester: John Wiley & Sons, Ltd.
6. National Center for Biotechnology Information. http://www.ncbi.nlm. nih.gov/ (accessed May 10, 2010).
7. May Patrick, Ehrlich Hans-Christian, and Steinke Thomas. "ZIB structure prediction pipeline: Composing a complex biological workflow through web services." In *Euro-Par 2006 Parallel Processing*, edited by Wolfgang E. Nagel, Wolfgang V. Walter and Wolfgang Lehner, Berlin: Springer, 2006. 1148–-158.

10

Using RFID/NFC for Pervasive Serious Games

The PLUG Experience

ERIC GRESSIER-SOUDAN, ROMAIN PELLERIN, AND MICHEL SIMATIC

Contents

Introduction

PLUG stands for "Play Ubiquitous Games and play more." It was a 27-month research project that started in January 2008. It focuses on pervasive multiplayer serious games, with companies and academics joining their skills to provide new results. France Telecom-Orange is a network provider. Net Innovations is a software integrator, and TetraEdge is a game studio. DUNE Adventure, providing real games such as crime investigations and spying, is directing characters. CNAM-CEDRIC, L3i-LaRochelle, and Institut Telecom are research labs. The Musée des Arts et Métiers (MAM) is a museum of science and technology in Paris. PLUG games take place inside the MAM.

This project involved around 25 persons, was funded by the French National Research Agency (ANR, Agence Nationale de la Recherche), and was labeled by Cap Digital, a business, research, and development cluster from Paris area dedicated to video games and digital entertainments. It was a 410 K€ founded project for a total cost of 1560 K€. PLUG explores ubiquitous computing, affective computing, game design, player monitoring, affordance, and acceptance. It mixes game experience and learning to enhance visitors' knowledge about inventions, research, and sciences.

PLUG inherits from past pervasive games projects. In the field of mixed reality, the eponymous Human Pacman [CHE04] was a reference. PAC-LAN [RAS06] stated the relationship between pervasive games and RFID/NFC tags. Via Mineralia [HEU07] in the Museum Terra Mineralia of Friberg combined RFID/NFC and serious games. The latter asserted the choice of using RFID/NFC technology to access contents in the context of a museum, but it didn't fully use the abilities of RFID/NFC that are not only read but also written as we do in PLUG. *REXplorer* [BAL06] enabled the vision of a social game in the context of a patrimonial tour. *Relax to Win* [SHA03] established the link between games and biofeedback leading to affective computing. Finally, the goals of the *IPerG* European project [IPE08], including the magic circle approach [HUI49; MON05], motivated us to propose our own pervasive games. Finally, PLUG also inherits from the *pussyfoot* invented by science fiction novelists [POH69] and which anticipated today's smart phone.

At the same time, PLUG was not our first ubiquitous game. We first worked on mobile games [PEL05]. We learned that wide acceptance was a key issue in the evaluation process of such new games [SIM04]. We also implemented SoundPark [PEL09a], a 3D audio mixed reality game, played in the Jane Mance Park in Montreal, that use RFID/NFC tags as clues to find virtual sounds. Museum context and pervasive games need to focus not only on game design, but also on urban design [GEN08]. Urban design encompasses buildings, objects, and the atmosphere inside the museum. They are of significant importance to providing take-up by players and visitors. Finally, affective computing allows basic functions like checking museum social constraints, for example, "It is forbidden to run inside the museum." Through players' emotional states, it opens new game design features. It leads to monitor players/visitors' behavior. Then, analysis of the way players follow the game scenario can be done. Finally, live adaptation of the game scenario is achieved when necessary [CHA05].

RFID/NFC technology combined with mobile phones provides content very easily. It is affordable. Young people and their families enjoy discovering artifacts this way. They are able to share a unique and immersive experience while visiting a museum. The aim of this chapter is to describe our experiments with RFID/NFC technology applied to pervasive multiplayer serious games. During the project two games have been implemented: PLUG: Secrets of the Museum (PSM), and PLUG: The Paris Overnight University (PPOU). Simplicity is characteristic of the technical architecture of PSM. It uses intensively passive RFID/NFC tags, and reads/writes tags through Nokia 6131 NFC mobile phones. No data communication is required at all. Players can talk with each other using voice communication, but this function is not central. The game is fun to play and educational. It brings a new kind of social interactions between players/visitors [JUT09]. PPOU is different. It is knowledge-centric, and the game design is built around learning. It is also more sophisticated. It makes full use of sensors and sensor networks, mixed reality, player monitoring, Nokia 6131 NFC, RFID/NFC tags, iPhone, ubiquitous middleware, and scenario adaptation. It is also based on nonplayer characters that act during the game session to foster immersion and to enrich the gameplay. The two games have been played during the museum open hours. As a consequence, game sessions occurred with the public unaware. It was interesting to watch how the public tried to enter the game and to analyze their interest in mimicking PSM or trying to be involved in object questing from PPOU. But this is the purpose of a different work related to social aspects of the PLUG project.

The chapter is organized as follow. The section titled "PLUG the Secret of the Museum" describes the game PSM. It presents the game design, the architecture, related works and outcomes. The section titled "PLUG the Paris Overnight University" presents PPOU, following the same organization. The section titled "Conclusions and Future Work" concludes the chapter and provides perspectives on the future.

Plug: Secrets of the Museum

PSM provides a new way to use RFID/NFC technology. There is no centralized server. No data communications are required. We provide a description of the game and key features of its implementation in this section.

Game Design

Game Stuff   Eight teams are involved in a PSM game session. A team is made up of two players. Each team is given a handset able to read/write RFID/NFC tags. We use Nokia 6131 NFC mobile phones. Players handle virtual cards. Cards are associated with meaningful objects/people of the museum content. Cards contain two types of information: the picture of an object from the museum, and the family the object belongs to. For example, the *Ghost Busters* family puts together famous scientists of the museum. Each handset carries four cards as depicted in Figure 10.1. Moreover, 16 passive RFID/NFC tags are spread throughout the museum. Each one contains one virtual card. The RFID/NFC tags are made visible to visitors and players, and they are close to meaningful objects of the museum as depicted in Figure 10.2. There are four families with four cards each. To reduce risks of deadlocks between players, every card is replicated; three copies are spread in the game. Finally 48 virtual cards are shuffled and distributed between eight handsets (four cards per handset) and the 16 RFID/NFC tags (one card per tag).

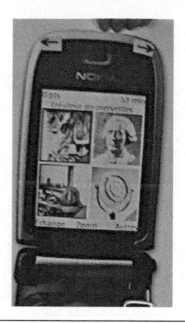

Figure 10.1 Four virtual cards carried by a Nokia 6131 NFC.

Goals The goals are twofold. The goal of the museum is to make players learn content. The goal of teams is to get the best score throughout a game session. These goals are achieved through the game design of PSM, and especially through the gameplay.

To gain points, each team may prove four skills:

- Searcher—by gathering on its handset four cards of the same family. It can be done through exchanges as shown in Figure 10.3:
  - "card stored in their handset ⇔ card stored in a RFID/NFC tag."
  - "card stored in their handset ⇔ card stored in another team's handset."

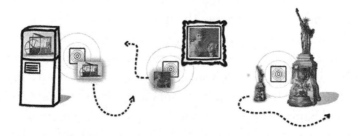

Figure 10.2 Association between meaningful artifacts and RFID/NFC tags at the museum.

Figure 10.3   NFC reading phone to tag or phone to phone.

- Custody and method—by bringing back a card in the RFID/NFC tag close to the object of the museum it refers to. This is achieved by an exchange "card stored in the handset ⇔ card stored in a RFID/NFC tag." This RFID/NFC tag is called home tag for the card.
- Generosity—by exchanging one of their four cards with another team.
- Curiosity—by answering quiz related to objects of the museum.

Gameplay   Once the game is set up, each team walks throughout the museum in order to interact with tags. To do so, a team positions its phone over a RFID/NFC tag, in order to read its contents thanks to the NFC ability of the Nokia. The handset displays the virtual card stored in the tag, an *Indication menu* item, and possibly a *Quiz menu* item. The team can exchange this virtual card with one of the cards of its handset. If a card is transferred (written) towards its home RFID/NFC tag, the team gains points for its custody and method as mentioned before. If the team selects *Indication menu* item, handset displays the location of a RFID/NFC tag which contains an object of the family gathered by the team, and the age of this information. Thus, the team can estimate the freshness of this information and if it is worth taking it into account. *Quiz menu* item is displayed only if the target RFID/NFC tag is the home of the card it stores or of one of the cards held by the player. When the team selects Quiz menu item, the handset displays a question (randomly selected among a set of questions) dealing with the object. If the team answers correctly, it gains points for its curiosity. If the team does not answer correctly, the handset displays an error message indicating the good answer and the team does not score any points. To be able to answer the question correctly, the team can lean on its members' knowledge, but it can also read museum's information. At game time, a team can also exchange cards with another team. To do so, both teams position their handsets very close to each other; the exchange is achieved in peer-to-peer NFC mode. Each concerned team receives points for its generosity.

Game session ends after one hour. Each handset records the points granted while collecting cards from its assigned family. It adds these points to the points gained for skills during the session. Then it displays the score of the team. The winner is the team with the highest score.

Benefits   PSM is educational. First, each team can "zoom" on objects held by its handset to get textual information on it. Moreover, virtual cards "speak," providing additional humorous information on the object. Finally, the players have to go through the whole museum in order to maximize their score. They discover pieces of art of the museum they would not have paid attention to if they had not played PSM. In addition, during the game, players are roaming around the museum to collect points. They do not follow the linear path of a regular visit. This feature of the game fosters the ability to memorize the content of the museum. So PSM is certainly educational. Nevertheless, it has been judged to be very entertaining by the players who tested it (around 700 people).

Because there are several ways of gaining points, PSM reaches a broad audience. Some players, especially young children, are mostly interested in the collector aspect of the game. Some other players, especially teenagers, mostly focus on exchanges and strategies to prevent other players from filling their collection. Other players, *a priori* juniors and seniors, mostly focus on the quiz aspect of the game. This game suits the requirements of a family visit. Family members can choose to play together in the same team. But family members can also play against each other, each member owning her own handset. Whatever their style (cooperating or challenging), every member of the family has fun playing this game while enriching their knowledge.

Installation of PSM is simple. The only thing to do is to deploy tags throughout the museum. Moreover, no network is needed at game time. Thus installation and operating costs are limited.

Architecture

Overview   PSM is built on top of a causal distributed memory (DM)) based on RFID/NFC. It fully uses vector clocks [FID88, MAT88] associated with RFID/NFC tags. This association may seem peculiar; we associate vector clock and versioning. Usually an element of a vector clock is associated with a processor. In our case, it is associated with a chunk of memory. Distributed causal memories, defined in [RAN95], use the same mechanism. But the combination of caching and gossiping with RFID/NFC tags and mobile handsets is brand new, as far as we know. Full details of the architecture are given in [SIM09].

Tags and handsets propagate information related to the game, while players exchange cards, and they cache the value of cards. The content of tags and handsets reflects game states as a true distributed shared memory. Vector clocks provide causal ordering of data propagation. Thus, handsets are able to read the cached value of any chunk of the DM without being physically close to the tag hosting that particular chunk. We present how vector clocks are used to give to each handset the best up-to-date view of the DM and thus to allow users to make queries on a local view of DM. Here are the main building blocks of the solution:

Main Principles   Each tag and mobile handset of the system holds a local view of DM. A vector clock is associated per local view.

Whenever a mobile handset comes close to a tag or another handset, the two elements involved build their own consistent view of DM by comparing their vector clock values. Doing so, they get a more up-to-date view of DM. An overview of the DM architecture with handsets and tags is given in Figure 10.4.

The system we consider is made of two types of components: RFID/NFC tags and mobile handsets (See Figure 10.4). Each element, e, holds a local view, DM_e, of the distributed memory. We note $DM_e[r]$ the view element e has of the contents of DM held by the RFID/NFC tag r. Each element e holds also a vector clock, VC_e, which is used to propagate operations done on DM [SAI05]. In the implementation, to save space on each tag, $VC_r[r]$ holds the timestamp of the last update done on $DM_r[r]$. Remember that it is the part of the distributed memory DM held by RFID/NFC tag r, whereas $VC_{e,e\neq r}[r]$ holds the timestamp of the last update of $DM_r[r]$, which element e is aware of. At initialization time, for all elements e of the system, DM_e is initialized with the initial value of the distributed memory. On the contrary, VC_e is initialized to a value independent from the application: $(0, \ldots, 0)$. Then, each time a mobile handset m changes the value stored in $DM_r[r]$ of a tag r, it applies Algorithm 1.

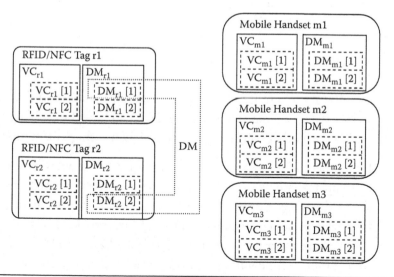

Figure 10.4 Combination of vector clocks and cached cards in a system made up of two RFID/NFC tags and three mobile handsets.

Algorithm 1: Update of $DM_r[r]$ on tag r by mobile handset m

1. $DM_r[r]$ $\leftarrow$ update of $DM_r[r]$
2. $VC_r[r]$ $\leftarrow$ date of update of $VC_r[r]$
3. $DM_m[r]$ $\leftarrow DM_r[r]$
4. $VC_m[r]$ $\leftarrow VC_r[r]$

Moreover, by applying Algorithm 2, DM_e and VC_e may be updated whenever element e is able to exchange information with another element e'. In the context of an RFID/NFC-based application, this happens in two cases: A mobile handset is near a RFID/NFC tag (that is, another mobile handset) and is able to interact with it through the RFID/NFC protocol (that is, an NFC peer-to-peer protocol).

Algorithm 2: Making DM_e and $DM_{e'}$ consistent

1. foreach i, $1 \leq i \leq$ number of tags in the system
2. if $VC_e[i] < VC_{e'}[i]$ then
3. // Element e' holds a more up-to-date view of $DM_i[i]$
4. $DM_e[i]$ $\leftarrow DM_{e'}[i]$
5. $VC_e[i]$ $\leftarrow VC_{e'}[i]$
6. elseif $VC_e[i] > VC_{e'}[i]$ then
7. // Element e holds a more up-to-date view of $DM_i[i]$
8. $DM_{e'}[i]$ $\leftarrow DM_e[i]$
9. $VC_{e'}[i]$ $\leftarrow VC_e[i]$
10. endif
11. endforeach

Liveness Thanks to our use of vector clocks, each element e of the system has the best up-to-date view of the DM it can have, by exchanging information with other elements. But this view has three limitations:

- First, there is no guarantee that this up-to-date view is the most up-to-date view of the DM. Due to players' interactions the DM evolves. So there may be an element e, e ≠ r for which $DM_e[r]$ does not correspond to $DM_r[r]$ currently held by tag r. But we have the guarantee that tag r did hold value $DM_e[r]$ at some point in time (either it is the initial value, or it is a subsequent value that induced the update of VCr[r] (Algorithm

1) and thus the propagation of this new value to element e (Algorithm 2).

- Second limitation: Element e has no guarantee that its global view DM_e corresponds to a value of DM that did really exist at some point in time [FID88, MAT88]. But causal ordering of events is ensured.

- Final limitation: Element e has no guarantee that it sees the whole history of changes of $DM_r[r]$ (for any tag r of the distributed system). Suppose that, at time t1, a mobile handset m1 sets the value of $DM_r[r]$ of tag r to v_{t1}. Algorithm 1 provides $VC_{m1} \leftarrow t1$. And, at time t2, a mobile handset m2 sets $DM_r[r]$ to v_{t2}. Algorithm 1 provides $VC_{m2} \leftarrow t2$. Afterwards, if m2 reaches tag r' before m1, Algorithm 2 provides $DM_{r'}[r] \leftarrow v_{t2}$ and $VC_{r'}[r] \leftarrow t2$. When m1 comes to tag r', because $VC_{m1}[r] < VC_{r'}[r]$, $DM_{r'}[r]$ is never set to the value v_{t1}.

Despite these limitations, an application running on a mobile handset, m, is able to make queries on DM_m; it has a consistent view of the whole distributed memory without moving physically towards each tag. Notice that the application can help reduce the negative effects of the limitations. As in gossip protocols, if it stimulates information exchanges between elements of the system, propagation of updates will disseminate quicker in the system. Thus, the gap between DM_e and DM will be thinner.

Implementation Hints and Runtime The game runs on the Nokia 6131 NFC. It is implemented as J2ME midlets. RFID/NFC tags characteristics are ISO 14443, Mifare-NFC, 13.56 MHz, 1 KB of RAM.

DM_e (that is, VC_e) is initialized for the 16 card values held initially by the tags (respectively $(0, \ldots, 0)$). At play time, whenever a team wants to exchange one of the cards held by its mobile phone with the card held by a tag, it must go close to the tag and the application on the mobile handset applies Algorithm 1. Moreover, when a team wants to know what card is physically contained in a tag, it also has to go nearby the tag. We take advantage of this read operation to apply Algorithm 2 in order to make DM_{mobile} and DM_{tag} consistent. When a team accesses a tag, a read operation occurs to display the card stored in the tag; there is also a write operation that modifies DM_{tag} and VC_{tag}, if necessary. When two teams exchange cards between their mobiles phones m1 and m2, Algorithm 2 is applied through the NFC peer-to-peer protocol: $DM_{mobile1}$ and $DM_{mobile2}$ are updated and made consistent. As a consequence, interactions from players make the DM alive, and update the causal history of accesses to tags. The game design is built to foster users' interactions. For example, players gain points for "custody and method" or for "generosity."

To help players in their quest, game design introduces a hint function. A team can ask for an indication where to get a virtual card that matches his collection. The user will be driven towards the tag that is supposed to store the proposed virtual card. The hint function is supported by an analysis of DM_{mobile} and VC_{mobile}. This function considers the virtual cards stored in DM_{mobile} that correspond to the family collected by the team. Among these, the process selects a card the team misses and for which the information known through VC_{mobile} is the most recent from the causal ordering of an events point of view. This ensures the update of the virtual card from the closest past in the DM_{mobile}.

Teams gain points when they exchange a card with another team. The lazy swap syndrome could occur if two teams were spending their time exchanging cards without moving to get cards in the museum. To prevent the lazy swap syndrome, a rule has been provided in the game design. A team can exchange at most two cards with another team every 10 minutes but no less. This rule limits the number of swaps made by a team during a PSM session. For example, in a session of an hour, a team can only make 12 swaps. It is not sufficient to win. This rule has a positive influence on the dissemination process of data inside the DM. Teams have to move to tag objects provisioning points. Thus, the liveness of our causal DM is enforced.

Related Work

As mentioned before, the work presented in this section relies on vector clocks [FID88, MAT88]. It mixes vector clocks and memory chunk versioning [RAN95] but in a ubiquitous application. Our protocol can be viewed as a gossip protocol [BIR07].

Reference COU09 presents an RFID/NFC-based distributed memory that does not require any global network. This distributed memory is illustrated through two applications: Ubi-Check and Ro-boswarm. In Ubi-Check, an RFID/NFC tag is attached to each of the traveler's items. Each tag is initialized with a value specific to the traveler. All of these RFID/NFC tags are read at special spots (for instance, after an airport security control). Their values are transmitted to an application that checks if they are consistent. If that is not the case, it means that, at some point, the traveler exchanged one of his items with the item of another traveler. An alarm is thus triggered to warn the traveler that one of his items is missing. In Robotswarn, RFID/NFC tags are placed throughout a physical space to give direction information to plain robots roaming around in this space. These tags are initialized by dedicated robots before regular robots are able to run. The major difference from the architecture described before is that, in both applications, distributed memory data cannot be modified any more once the initialization process is over. In other words, Reference COU09 introduces an RFID/NFC-based distributed ROM that can be "flashed" (initialized), while we introduce an RFID/NFC-based distributed RAM.

As PSM, Save the Princess! is a pervasive game implementing an architecture where tags are read/written [MOT06]. It is based on motes. Similar to PSM, this game does not need any global network. When a mobile comes near a mote, TinyLIME allows access to the information stored on that mote. But, TinyLIME does not provide a means to get a simultaneous access to the overall information stored in all the motes as PSM does through its gossip protocol and causal consistent data management framework.

Outcomes

PSM has been played by around 700 people, most of them are kids. There were 250 kids who tried the game during two days, February 23 and 24, 2010. It is more than a simple proof of concept. Despite the end of the project, PSM is still used inside MAM. Kids enjoy playing and discovering the content of the museum this way. It is very attractive. PSM is completely handled now by people from the museum without any help of the developer team. From our point of view, PSM has been very successful.

The game has been evaluated by people from museums and sociologists. PSM is a teasing for the MAM, but its strength is also its weakness. It is difficult to find the right balance between play and learn. Nobody can tell that learning cannot be achieved through PSM. But it does not satisfy completely the educational goal required from the museum point of view. Kids, or more generally visitors, play more than they learn. The game design closed the magic circle [HUI49; MON05]. We made this observation after the end of the project, and it will be taken into account in any future projects.

We tried to adopt a new approach more learning-centric in the second game that the PLUG project implemented. New guidelines have been proposed to design a new game. The next section provides details on this second game and its main features.

PLUG: The Paris Overnight University

The game PLUG: The Paris Overnight University (PPOU) is also a RFID/NFC-based architecture but not only. RFID/NFCs are the basis of artifact identification and localization. The players' access tags through NFC readers, Nokia 6131 NFC mobile handsets. From a technological point of view, the new game is middleware-centric. It is based on the uGASP middleware [PEL08]. uGASP offers an easy way to deliver content associated with RFID/NFC tags to players on their iPhones. It implements the model-view-controller design pattern [GAM95]. RFID/NFC tags provide information. Data are relayed toward the game logic implemented on the uGASP server through 3G mobile communications. iPhones support rendering of text, images, and short videos.

Game Design

Game Stuff The complexity of PPOU is greater. RFID/NFC is a key component to identify artifacts, but the game design also uses sensors, sensor networks, player behavior monitoring, handsets, a server, mobile phone communications, network management, and Wi-Fi communications. These components support the game play and help to deliver an immersive experience to players while they learn about MAM content.

- **Indoor location handling though device management**: An SNMP (Simple Network Management Protocol)-based network management architecture has been deployed. This architecture relies on embedded computers and Wi-Fi communications. The museum is entirely covered using Wi-Fi access points. Each access point defines an area. This communication infrastructure provides hints about where non- player characters or people from the staff are. This localization is made possible thanks to a tiny computer, a Gumstix, they carry. Each computer is tracked on the field and not geolocalized.

  This service did not work as well as we expected. Investigations showed that the embedded processor was not powerful enough. It was also unable to support smooth roaming between Wi-Fi areas. Despite many attempts, we did not succeed in finding the right configuration that was reliable enough, with batteries that would last until the end of a game session.

- **Sensor networks and sensors [DUP08]**: During the game, teams compete to get more points. The battle mode allows results to be stolen from a concurrent team. The process is very simple. Each team provides a thief. It wears a heart rate sensor. The thief targets another team. The targeted team has to defend itself through its thief. To do so, the player with the heart rate sensor has to change his heart rate to be close to that of the thief. If he succeeds quickly enough, he will be able to pick up the results held by the thief. Otherwise, he is stolen and the thief succeeds.

  Measurements of heart rate sensors from each team are gathered to a central database through a ZigBee network that meshes the museum area where PPOU takes place.

- **RFID/NFC reading through Nokia 6131 NFC handset**: Each team has one Nokia handset. This device is used only to read tags (and not to write them as in the previous game). The content of tags is sent to the game logic, which is running on a uGASP server. It is used to locate players inside the museum just when they read tags, and to provide identification of rooms and objects from the game where the team is playing.

- **iPhones as interaction devices**: Each team has one iPhone. The iPhone functions as a schoolbag, and its content can be stolen (hacked) by other students. iPhones are used to support interactions with players—to provide information on the object of the game, to run quizes, to get answers from teams, and to show results following the PPOU game logic.
- **Players monitoring [CHA05]**: Each time a team does something in the game, the corresponding information is used to make progress in the game, but it is also reported to the monitoring server, and to the game masters. This continuous reporting of meaningful game actions is used to help players to decrease difficulties, to slow down other players, and to help people blocked in the game. This tool is also very helpful to get feedback on how people play, that is, the efficiency of their game design.

The following sections consider gameplay and how RFID/NFC tags are involved in the game. The description focuses on uGASP.

Goals The museum contains meaningful objects. The goal of the game is to discover through different objects of the museum why and how the targeted object has been invented, what are its main characteristics, who provided it, etc. Players are organized in teams of two to six persons. We tried different combinations of these numbers in a team. Each configuration leads to a different gameplay with different social interactions. It seems that three offer the best configuration: one player hunts RIFD/NFC tags to validate rooms and objects to be discovered, the other two answer questions on the iPhone, and are able to compete during student challenges or content hacks.

During our experiment, the targeted object was the Hydroptère [HYD10], a sailing boat able to "fly" over the ocean at a very high speed. Some of the intermediary objects were Clement Ader's plane [WIK10], the computer, composite materials, and building frames. To validate intermediary objects that they cannot figure out without following instructions on the iPhone, players tag rooms, get information on video/pictures/web, answer questions, and try to tag the right object in the room. At the end, what players have learned is verified through questions. The way answers are provided is highlighted in the next section.

Gameplay Players are invited to join an investigation. Police have discovered some strange activity in the museum, something that occurs in parallel and which is not official. Players become agents under cover. They have to infiltrate an organization, Paris Overnight University (POU), and report what is going on.

To do so, they are recruited as potential students of POU. The police lieutenant, dressed like a museum guide, introduces the players through a tiny door in a darker part of the museum. The game starts. They meet Gilles Picquart, the student advisor, and Big Z, the dean of the University. They learn that they have to prepare an exam to be admitted into POU. The exam is related to the content of the museum and then of the game. The iPhone is their schoolbag and they need to fill it correctly, step after step. When the game is completed, they have to go to an unexpected room in the museum to pass the admittance examination. The jury is composed of Augustine Meulard, the director of the university, whom they have never seen before, the student advisor, and Big Z.

The game ends by the examination. All the students/players we had passed it. Before leaving, they need to swear that they will not reveal the POU secret organization. When they go out of the exam all together, the police lieutenant asks what POU stands for and what it is about. A few reveal the secret, but some do not. Anyway, the players enjoyed it very much and had fun.

The game design, because learning is a key issue, is closer to the flow approach from [CSI90] than to the magic circle approach [HUI49].

Architecture

Overview The game design described in the previous section featured content as the main goal. Support for the various interactions mandates the use of several technologies from RFID/NFC reading to multimedia content delivery on iPhone through player traceability with game masters. Integrating these components together is the aim of a middleware able to handle ubiquitous computing applications.

We used uGASP [PEL09b]. uGASP stands for ubiquitous GAming Services Platform [OW210]. It is the foundation of our solution. uGASP also provides the game engine that maintains game states and achieves game logic in a highly mobile and interactive context. Network and device management as affective computing have been handled in PPOU as autonomous building blocks. This section focuses on uGASP and the way it integrates RFID/NFC services.

Main Principles uGASP implements the Open Mobile Alliance Gaming Services (OMA GS) working group specifications that deal with multiplayer game management on embedded devices. uGASP extends OMA GS proposal with services for ubiquitous gaming. uGASP is a service- and component-oriented architecture. Its services include network communication, session management, a game server engine that handles game logic, ubiquitous services as geo-localization or devices management, and system services. Figure 10.5 gives an overview of the full set of services currently provided by uGASP. The architecture has been improved in the context of the SoundPark project [PEL09a]. It is an ongoing project. The next step will bring uGASP towards the Cloud.

Existing components developed for previous ubiquitous games [PEL09b] have been extensively used on the server and client implementations of PPOU. The RFID/NFC handler was deployed on mobile phones and the corresponding OSGi bundle on the server.

The client/server communication uses an object-oriented protocol called MooDS (Mobile optimized objects Description and Serialization), a key component of uGASP [PEL07]. It is an abstract syntax notation and a transfer syntax to exchange objects of a mobile game. This protocol has been created specifically due to address multiplayer games development and mobile network requirements. It reduces a significant volume of transmitted data, increases communication speed, and lowers the memory footprint of the implementation. It has been compared to J2ME SOAP (Simple Object Access Protocol)-based protocols, and its efficiency has been proven [PEL07]. It is a key building block of uGASP.

The middleware has been also improved with a bidirectional network communication connector, socket based and MooDS conformant, supporting more efficient exchanges than available with the HTTP connector.

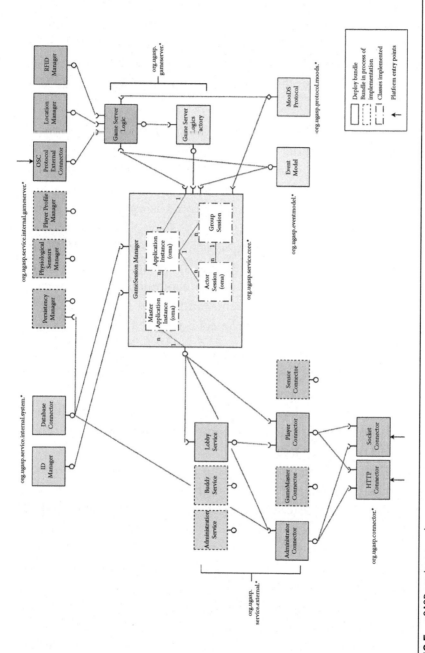

Figure 10.5 uGASP services overview.

Implementation Hints and Runtime   uGASP is implemented in JavaME and Java. It is based on the Open Services Gateway initiative (OSGi) component framework, and more precisely on the iPOJO-OSGi layer [OSG10] that provides additional modular, dynamic, and configurable services through the creation, deployment, and calling of bundles. With iPOJO-OSGi, the designer of the solution has to specify the uGASP services he needs, and the specific application components he will use:

- The required uGASP services are specified as components with their required and offered interfaces. These are then instantiated automatically as appropriate during initialization of a game session on the server side.
- The game logic engine is implemented as a Java class within the server.
- The component-oriented approach inherited from iPOJO-OSGi provides a guideline to specify the specific application bundles, the data model, and the communications between bundles. Finally, the different bundles can be implemented separately, by different developers if required.

Due to the iPOJO-OSGi execution platform, the deployed byte-code in the overall PPOU game contained only the necessary bundles, providing an optimized application memory footprint, as required for resource-constrained devices. An extension of MooDS has also been implemented in C++ to support communications between the iPhone (C++) and the server (Java).

Finally, two ubiquitous computing architectures have been tried during the design phase of the game PPOU. The first one relied on a Wi-Fi network. Instead of iPhone we used iPodTouch handsets. Nokia 6131 NFC were connected to the wireless network through Bluetooth and a tiny embedded processor, a Gumstix, able to multiplex sensor communications over Wi-Fi. This first architecture failed: roaming under full experiment was inefficient, and battery consumption was huge (no more than one hour autonomy for Gumstix and for iPodTouchs). Due to project timing we were unable to spend much time on these problems. Then, 24 hours later, the project switched to a new solution. It was based on 3.5G mobile phone communication.

iPodTouchs were replaced by iPhones instantaneously. Nokias and iPhone communicated directly with the uGASP server. The new solution worked immediately, and we based further experiments on it.

Related Work

Pac-LAN RFID/NFC [RAS06] is a passive RFID/NFC-based game. It is inspired by the well-known PacMan arcade video game. The tags represent resources of the game, mainly pills that are geo-localized resources. Pac-LAN runs on Nokia mobile phones, the ancestor of the 6131 NFC we used in PLUG and is implemented in JavaME also.

As far as we know, the only game that provides the same overall features as PPOU is Momentum from the IPerG European project [IPE08]. Momentum is "more": longer (3 weeks), larger (many more players), more immersive (it is a role-playing game in a city), and also more expensive. But Momentum is not museum-related, nor content-centric. Its game design perfectly matches the magic circle approach [HUI49]. It uses equivalent technologies but goes further toward immersion—for example, technology is embedded in artifacts. The RFID/NFC reader is embedded in a glove, the Thumin Glove [IPE07]. IPerG, and especially Momentum, is an amazing project dedicated to pervasive gaming.

SoundPark [PEL09a] is a mixed reality game. The virtual world is 3D audio. Clues to discover sounds from music instruments in the virtual world, such as piano, drums, and djembe, are provided by passive RFID/NFC tags. It is supported by uGASP, and its architecture was used to prepare PPOU. It implements the MVC pattern based on RFID/NFC tags.

ReXplorer [BAL06] is close to PPOU because it is content-centric. It fosters the discovery of the Regensburg town as a Middle Ages adventure, but it does not use RFID/NFCs.

Outcomes

PPOU has not been played as much as PSM. Taking into account the preliminary trials, it has been played by around 100 players. The final demonstration has been tried by 30 players, including evaluators of the project and people from different museums of Paris like the Louvre, Musée du Quai Branly, and Cité des Sciences de La Villette. People enjoyed it, and some of the evaluators played both games. They confirmed that PPOU is more learning-centric. Our goals were successful. Today, PPOU is contributing to a new project based on participative design, where students from secondary school or high school build the content of the game itself. At the end, PSM and PPOU provide two ways to discover a museum. We consider them to be complementary approaches.

In PPOU, we used RFID/NFC in a very classical way; we only read RFIDs. But effectiveness of RFID/NFC technology to support mixed reality serious games has been asserted. It is easy to use. It is a flexible means to provide hints toward content. And finally, its use is not limited to Electronic Product Code global-based architectures.

From the ubiquitous computing point of view, we have to go back to the Wi-Fi–based solution to discover why it failed. We are convinced that a wireless area network is more suitable for the museum because it lowers the cost of the overall solution. We could use iPodTouchs extended with an RFID/NFC reader or any Wi-Fi–based smart device with this facility. We expect that RFID/NFC-enabled smart devices will appear soon. To lower costs, museums need to use on-the-shelf products. For example, the San Francisco Museum of Modern Art replaced its multimedia guides by iPodTouch handsets [PRO09].

Conclusion and Future Work

Two kinds of solutions have been tried:

- PSM provided a distributed system approach with simplicity as the main requirement on the architecture. Neither Wi-Fi nor mobile phone communications are needed. People of the museum have been able to run it without specific knowledge of the technology involved. For instance, 250 children from primary school played for two days, one group after another. Today, around 700 people played PSM without any significant problem due to the technology. The solution has been improved by experiments in the field. The game can be performed by any RFID/NFC-enabled device. It can be played through sessions or continuously during daytime. The next scientific issue is to address scaling issues.

- PPOU provided a client/server approach, based on the model-view-controller design pattern, with a ubiquitous middleware that fully relies on mobile phone communications. The game can be played as event sessions. It is more complex and needs engineering support. We did not have enough opportunities to test the game because the project ended soon after we delivered it, and sessions are heavy to schedule. But we noticed that there are ways to alleviate this drawback. The game can be extended to more teams and should be able to scale due to the client/server approach. The game could be reduced to the quest toward a representative meaningful artifact. This would enforce learning but would also lower gaming. It could be played continuously during daytime.

This chapter does not deal with games evaluation issues such as acceptance, affordance, playability, etc. But the PLUG project, and especially partners from social science, and from the museum, spent a lot of attention and time on these aspects [JUT09, DAM10]. This step is mandatory in the process to provide a real solution that suits public requirements.

The two games we have presented here showed the effectiveness of RFIDs combined with NFC to build ubiquitous, mixed-reality-based, serious games. NFC eased and then fostered interactivity. Players figured immediately how to use it.

If we had to summarize the contribution of NFC in PLUG, we will choose the "digital kiss," as shown in Figure 10.3 (right side). It was an unexpected feature of PSM discovered by players themselves when they exchanged data from handset to handset, handsets vibrating during the transfer. The technology brought the magic of the digital world to players, enabling them to achieve immersion in a mixed reality world.

References

[BAL06] R. Ballagas, S. Walz, and J. Borchers. REXplorer: A pervasive spell-casting game for tourists as social software. CHI'2006 Workshop on Mobile Social Software (MoSoSo). April 2006.

[BIR07] K. Birman. The promise, and limitations, of gossip protocols. *SIGOPS Oper. Syst. Rev.*, 41(5):8–13, 2007.

[CHA05] R. Champagnat, A. Prigent, and P. Estraillier. Scenario building based on formal methods and adaptative execution. ISAGA'2005. Int. Simulation and Gaming Association. July 2005, Atlanta.

[CHE04] A. D. Cheok, K. H. Goh, W. Liu, F. Farbiz, S.W. Fong, S. L. Teo, Y. Li, and X. Yang. Human pacman: A mobile, wide-area entertainment system based on physical, social, and ubiquitous computing. *Personal Ubiquitous Comput.*, 8(2):71-81, 2004.

[COU09] P. Couderc and M. Banâtre. Beyond RFID: The ubiquitous near-field distributed memory. *ERCIM News*, (76):35–36, January 2009.

[CSI90] M. Csikszentmihalyi. *Flow: The Psychology of Optimal Experience.* London: Harper Perennial, 1990.

[DAM10] A. Damala, I. Astic, and C. Aunis. PLUG, Université Paris Nuit: A design reiteration of a mobile museum edutainment application. *The 11th International Symposium on Virtual Reality, Archaeology and Cultural Heritage VAST (2010).* Paris, France. September 21–24, 2010. To be published as a short paper.

[DUP08] J. Dupire, V. Gal, and A. Topol. Physiological player sensing: New interaction devices for video game. *Entertainment Computing— ICEC'2008: 7th International Conference*, Pittsburgh, PA, September 25–27, 2008.

[FID88] C. J. Fidge. Timestamps in message-passing systems that preserve the partial ordering. In *Proceedings of the 11th Australian Computer Science Conference (ACSC'88)*, pages 56–66. K. Raymond, February 1988.

[GAM95] E. Gamma, R. Helm, R. Johnson, and J. Vlissides. *Design Patterns: Elements of Reusable Object-Oriented Software.* Reading, Addison-Wesley Professional Computing Series. 1995.

[GEN08] A. Gentes, E. Gressier-Soudan, and I. Réchiniac-Astic. "Das Unheimliche" of ubiquitous games for museum visitors: When media environment becomes real. WMEBR'2008. Université de Berne. Suisse, February 4–6 2008.

[HEU07] G. Heumer, F. Gommlich, B. Jung, and A. Muller. Via Mineralia—a pervasive museum exploration game, In *Proceedings of the 4th International Symposium on Pervasive Gaming Applications*, PerGames 2007, Salzbourg, Austria. 2007.

[HUI49] J. Huizinga. *Homo Ludens: A Study of the Play-Element in Culture.* London: Routledge. 1949.

[HYD10] l'Hydroptère. http://www.hydroptere.com/_en/default.php?page=49 &sspage=54#centre. August 2010.

[IPE07] IPerG (Integrated Project on Pervasive Gaming). WorkPackage WP11: ELARP. Deliverable D11.8 Appendix C: Momentum Evaluation Report. Jaakko Stenros, University of Tampere. Markus Montola, University of Tampere. Annika Waern, SICS. Staffan Jonsson, SICS. May 15. 2007. http://iperg.sics.se/Deliverables/D11.8-Appendix-C-Momentum-Evaluation-Report.pdf. Accessed August 2010.

[IPE08] IPerG (Integrated Project on Pervasive Gaming). http://iperg.sics.se/index.php, 04/28/09.

[JUT09] C. Jutant, A. Guyot, A. Gentes, and M. Simatic. RFID technology: Fostering human interaction. IADIS Game and Entertainment 2009.

[MAT88] F. Mattern. Virtual time and global states of distributed systems. In *Proc. Workshop on Parallel and Distributed Algorithms*, Chateau de Bonas, France, pages 215–226. Elsevier, October 1988.

[MOT06] L. Mottola, A. L. Murphy, and G. P. Picco. Pervasive games in a mote-enabled virtual world using tuple space middleware. In *NetGames'06: Proceedings of 5th ACM SIGCOMM Workshop on Network and System Support for Games*, pages 29–36. New York: ACM Press, 2006.

[MON05] M. Montola. Exploring the edge of the magic circle: Defining pervasive games. *DAC 2005 Conference*, December 1–3. IT University of Copenhagen, 2005.

[OSG10] OSGi Service platform release 4 specifications, http://www.osgi.org/Release4/HomePage. August 2010.

[OW210] OW2 forge, uGASP under the L-GPL license from http://gasp.ow2.org/ubiquitous-osgi-middleware.html. August 2010.

[PEL05] R. Pellerin, F. Delpiano, F. Duclos, E. Gressier-Soudan, and M. Simatic. GASP: An open source gaming service middleware dedicated to multiplayer games for J2ME based mobile phones. *7th International Conference on Computer Games CGAMES'05*, November, Angouleme, France, pp. 75–82, 2005.

[PEL07] R. Pellerin. The MooDS protocol: A J2ME object-oriented communication protocol. In *Mobility Conference 2007*, September 10–12, Singapore, pp. 8–15, 2007.

[PEL09a] R. Pellerin, N. Bouillot, T. Pietkiewicz, M. Wozniewski, Z. Settel, E. Gressier-Soudan, and J.R. Cooperstock. SoundPark: Towards highly collaborative game support in a ubiquitous computing architecture. *DAIS'2009*. June 9–12, 2009. Lisbon, Portugal.

[PEL09b] R. Pellerin, Contribution to multiplayer ubiquitous games engineering. PhD thesis. CNAM Paris. September 2009 in French.

[POH69] F. Pohl. *The Age of Pussyfoot*. October 1969. New York: Ballantine, 212 p

[PRO09] N. Proctor. MuseumMobile. Podcasts from Museum Mobile Wiki. 13/05/2009. http://itunes.apple.com/podcast/id307747990. Also Peter Samis on Mobile Multimedia. http://museummobile.info/archives/221.

[RAN95] J. Ranjit. Implementing and programming weakly consistent memories. PhD thesis. Georgia Institute of Technology. March 1995. Also available as GIT-CC-95-12 at GATECH.

[RAS06] O. Rashid, W. Bamford, P. Coulton, R. Edwards, and J. Scheible. PAC-LAN: Mixed-reality gaming with RFID-enabled mobile phones. *ACM Comput. Entertainment (CIE)*, Vol. 4, No. 4. October–December 2006.

[SAI05] Y. Saito and M. Shapiro. Optimistic replication. *ACM Comput. Surv.*, 37(1):42–81, 2005.

[SHA03] J. Sharry, M. Mc Dermott, and J. Condron. Relax to win: Treating children with anxiety problems with a biofeedback video game. *Eisteach: Journal of the Irish Association for Counselling and Psychotherapy*, 2(25), 22–25, 2003.

[SIM04] M. Simatic, S. Craipeau, A. Beugnard, S. Chabridon, M. Legout. and E. Gressier-Soudan. Technical and usage issues for mobile multi-player games. *5th International Conference on Computer Games: Artificial Intelligence, Design and Education, CGAID'04*, pp. 134-138, 2004.

[SIM09] M. Simatic. RFID-based replicated distributed memory for mobile applications. *Proceedings of the 1st International Conference on Mobile Computing, Applications, and Services (Mobicase 2009)*. San Diego, CA, ICST, October 2009.

[WIK10] Ader Avion III. http://en.wikipedia.org/wiki/Ader_Avion_III, August 2010.

11

Empirically Grounded Design of a Nutrition Tracking System for Patients with Eating Disorders

PHILIPP MENSCHNER, ANDREAS
PRINZ, AND JAN MARCO LEIMEISTER

Contents

Background and Introduction

Patients suffering from chronic diseases often have to cope with limitations and a reduced quality of life. In particular, ALS (amyotrophic lateral sclerosis) patients suffer from progressive paralysis. This accounts for instances of insufficient dietary intake and, in the long run, leads to unnoticed reduction of weight due to malnutrition (Meyer, 2009), problems that are also found in other diseases such as dementia, Parkinson's disease or multiple sclerosis. From a certain point in time, enteral nutrition by use of feeding tubes becomes inevitable, which causes a dramatic decrease in patients' quality of life

(Löser et al., 2007). By maintaining adequate nutrition, especially in the early phases of the affliction, this process can be slowed down. For this purpose, due to the limitations caused by the disease, patients admittedly need medical care and the attendance of physicians, as well as the support of family members, affiliates, and nursing staff. What we advocate in this chapter, however, requires enhanced information logistics that are presently not available.

By the use of mobile networks and information systems, the insufficient information logistics among physicians, patients, and nursing staff concerning actual nutrition could be improved. Therefore, the objective of our work is to provide simple, effective, and efficient self-management of the current nutrition status. Furthermore, the potential solution would allow patients cost-efficient and easy-to-handle self-management of their current nutrition status and enable an autonomous lifestyle for a longer time span. Telecommunication networks allow flexible, location-independent monitoring of the current nutrition status of patients even in real time. Additionally, costs can be reduced, as patients take over data acquisition tasks. An intelligent nutrition tracking system thus can improve the productivity of medical processes not only by being more cost-efficient, but also by accounting for improved standards of medical care, as well as quality of life. This is rendered possible by the integration of patients into the treatment processes.

The presented case examines the situation of ALS patients. The outcomes are intended to be transferrable to other chronic diseases that encounter nutrition issues, such as multiple sclerosis or obesity.

Amyotrophic Lateral Sclerosis (ALS)

ALS is a progressive, degenerative motor neuron disease that, on average, leads to death within 3 to 5 years after diagnosis. It is considered to be a rare neurological disorder, and its origins are at present still unknown. In 100,000 people, approximately 6 to 8 people suffer from ALS, and each year another 2 develop the disease (Borasio and Pongratz, 1997). As both the lower and upper motor neurons degenerate, the disorder causes muscle weakness, and spasticity atrophy throughout the body. The consequences are indications of paralysis, among them a disturbance in swallowing or chewing (Cleveland and Rothstein, 2001).

According to current medical research, healing for ALS is unknown and inevitably leads to death. Therefore, actual treatments aim to control symptoms of ALS or can be considered palliative care.

In the course of ALS, undesired weight loss occurs due to malnutrition or cachexia (Cleveland and Rothstein, 2001). The loss of weight is accompanied by high morbidity and mortality, and leads to a decreased quality of life (Desport et al., 1999). Due to medical complications and social consequences, malnutrition and cachexia are of significant socioeconomic importance (Ludolph, 2006).

Supplementary nutrition by increased calories or the use of percutaneous endoscopic gastrostomy (PEG)-tubes for enteral nutrition is beyond that associated with significant effects of personal, logistical, and financial expenditures for service providers, as well as insurance and funding agencies. Estimates of the cost of malnutrition to the German public health and welfare system are about 17 billion euros (Löser et al., 2007). The annual costs for enteral nutrition for one patient is about 15,000 euros (Löser et al., 2007); considering the additional costs of complex care expenses, the total expenditures for a patient total approximately 50,000 euros per year (Schauder, 2006). Hence, innovations in ameliorating patients' nutritional status are relevant from both an economic perspective as well as quality of life.

Scenario

Several months ago, Hans was diagnosed with ALS, which started with a gait disturbance and progressive paralysis of his legs. During walking, he stumbles repeatedly, and he also encounters balance problems while going upstairs. Furthermore, swallowing increasingly causes him problems. Especially solid food (eating an apple), but also liquids are an absorption problem for him. His physician has created a nutrition plan where the daily caloric requirement is noted. Visits at the ALS outpatient department are scheduled every 3 months—too late to counteract early unnoticed weight loss. During the course of ALS, the PEG tube will become necessary at a particular time. The installation of the tube system can be delayed and the complication rate of the PEG-nutrition reduced if sufficient calorie intake prevents malnutrition. The physician has received a report from a new nutritional management system (currently researched and evaluated) with the simple possibility of recording and analyzing the daily food intake by information technology. If too few calories are ingested, the patient receives a warning. In the process, the caretakers and family members, as well as the physician will be informed. Although Hans has never been interested in technical equipment in his life, he can easily enter the data with the NFC-based nutritional management system. Everything is operated from his mobile phone: by touching the mobile phone with food images on a specially produced poster, the data will be sent to the system. Touching the poster twice is enough to send the data. Despite the process of the ALS-induced paralysis, Hans can independently record his daily nutrition by touching the food images on his mobile phone. With this continuous recording and the warning through the system, Hans has retained most of his original weight.

Related Work

NFC (Near Field Communication) is relatively novel in healthcare research. Especially for home healthcare solutions NFC is becoming more and more popular. Morak et al. (2007) used NFC technology for monitoring heart failure patients as a self-management process; Iglesias et al. (2009) describe an NFC-based health monitoring system to improve the quality of life for elderly patients. Patients transmit health-related data to a central database by touching medical devices with the mobile phone. Physicians or nurses can view the entire data and guide the patient to the best possible health status. Bravo et al. (2008) used NFC for supporting nurse activities in an Alzheimer's day center.

In the clinical context, NFC is used by various researchers. Lahtela et al. (2008) developed an NFC-based solution to avoid medication errors in hospitals. As an additional path of medical data acquisition, Fikry et al. (2006) and Morak et al. (2009) describe different NFC-based solutions that allow physicians or nurses to collect data by easily touching medical devices with a mobile phone.

Research Design

This section describes the key objectives of our research project, and the methods used to collect and analyze data, as well as how the research process will be implemented. The objective is to plan, build, introduce, and evaluate a nutrition tracking and management system based on mobile services and an IS platform. According to Schwabe and Krcmar (2000b), pilot projects are a special version of interventionistic science; they develop and implement technological innovations in their natural organizational and social environment.

The starting point is a socio-organizational problem (in this case the situation of ALS patients suffering from malnutrition). It begins with an in-depth analysis of the current situation of ALS patients. Therefore, a literature review and case studies, using interviews, questionnaires, observations, and document analyses (Yin, 1989), will be used. The perspectives on the research objects for analysis are deduced from the Needs Driven Approach (NDA) by Schwabe and Krcmar (1996). Originally, the NDA was developed to design tele-cooperation (Schwabe and Krcmar, 2000a), but it has already been successfully applied to other settings, for example, for the development of virtual communities for cancer patients (Leimeister et al., 2002). It analyzes tasks, work processes, interactions of actors, social structures, tools and shared material, adoption and diffusion of technology, and information storage (Schwabe and Krcmar, 1996). These perspectives are the basis for designing interview guidelines, analyzing documents in self-help groups, constructing questionnaires, and all other methods used in the phase of field studies. The results of the analysis are used for designing a nutrition tracking system. This system is then implemented in the field, and finally improvements in the system are made during the remainder of the project. At all times and on all levels a continuous evaluation takes place, and, in this way, iterative learning steps can be augmented at all stages. Thus, this pilot project can be considered a level-three pilot project, since it consists of analysis, design, and implementation of an information system (Schwabe and Krcmar, 2000b).

Requirements Analysis for Nutrition Tracking System

Analysis of Actual State and Processes

The objectives of this analysis are to evaluate ALS patients' needs and to elevate current information and interaction processes between physicians, patients, and nursing staff. Therefore, we analyzed the ALS patients' current situation by qualitative interviews and workshops with experts (physicians, nursing staff, and medical technicians), followed by workshops with ALS self-help groups consisting of patients and affiliates. To elevate current processes, observation and shadowing techniques were used. Based on these findings, target processes were formulated, which were also discussed and evaluated in workshops with experts and self-help groups. The following analysis is based on the perspective of the NDA. This method facilitates the transfer of the results into system development requirements.

ALS patients' current situation (interviews with patients, affiliates, physicians): Malnutrition is not only caused by swallowing disorders, but also by respiratory insufficiency, increased caloric needs due to hyper-metabolism, inability to use the upper extremities, or by depression. Additionally, dehydration often occurs. Thus, the nutritional state of the patient is an independent risk factor for survival (Desport et al., 1999). At a particular time, total enteral nutrition by use of PEG tubes is inevitable. In such cases, the quality of the patient's life decreases dramatically, and the mortality in the first months after installation of a PEG tube is increased (Forbes et al., 2004, Ludolph, 2006). Established treatment options include nutritional advice by aiming for hyper-caloric nutrition. In the case of swallowing disorders, use of logopedic measures or supplemental nutrition by use of specialized aliments is possible. If these measures cease to work, PEG tubes will be applied. Hence, the course of malnutrition encompasses three phases:

- *Phase 1—intensified nutrition*: The patient can still nourish himself, but already needs more calories than he actually ingests.
- *Phase 2—adapted nutrition*: The patient needs to enrich his nutriment with high-caloric products in order to ingest enough calories.
- *Phase 3—total enteral nutrition*: Without the use of PEG tubes, the patient cannot be nourished any more.

The progression of these phases is different from patient to patient. Due to increased risks and limitations aligned with phase 3, it is desirable for patients to remain autonomous as long as possible. Thus, securing and managing a sufficient and well-balanced nutrition is of the utmost importance in the earlier phases of ALS in order to delay total enteral nutrition. This can only be accomplished if nutrition and care processes are adopted efficiently and at the right time. Therefore, affiliates and nursing staff rely on information provided by physicians at the appropriate time.

Ethnographic analysis of treatment processes: Six observations of current treatments, consisting of shadowing and narrative interviews with physicians showed that there is a lack of information and interaction possibilities in general. The observations took place at an ALS-specialized clinic. Usually, patients show up there every 3 months to document the status of their disease and to adjust current treatment procedures. The physicians at this special clinic have no medical follow-up of their patients during the 3-month interval between visits; additionally, they do not receive any information from affiliates, nursing staff, or the family doctor. Therefore, the physicians need to rely completely on the information presented by the patient, which in turn has several major disadvantages: Patients fail to give the correct information (either done intentionally or due to inattention); information from nursing staff or the family doctor is sent via the patient; and sometimes due to the progressive paralysis, the patient has difficulties in communicating at all. Hence, the basis of decision-making is often an insufficient data set, which additionally consists of secondary data ("Chinese whisper effect" or time lag). Additionally, about 90% of the consultation time is used for information retrieval. At the end of the consultation, patients are given instructions and information intended for affiliated and nursing staff.

Social structure: The social structure describes who is interacting with whom and in which way. (See Figure 11.1.) Although there were some minor differences in the observed cases, a dominant pattern was identified: in the case of ALS patients, most interactions took place between the physician and the patient. Instructions and information intended for nursing staff or affiliates were sent via the patient. Correspondence between the physician and family doctor only took place, if at all, in the form of doctors' notes. In rare cases, affiliates did accompany patients during consultations. Care and assistance in patients' daily life were carried out by either affiliates or by professional nursing staff.

Processes and interactions: Figure 11.2 shows the treatment process of ALS in an ALS-specialized clinic as a service blueprint. Service

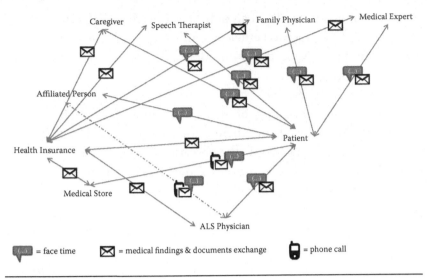

Figure 11.1   The typical interaction network of an ALS patient.

blueprinting is a process analysis methodology proposed by (Shostack, 1982, 1984). Zeithaml et al. (2006) define service blueprinting as a tool for simultaneously depicting the service process, the points of customer contact, and the evidence of the service from the customer's point of view.

Analysis of materials and tools: In the case of ALS patients, the use of materials and tools, such as mobile devices, seemed to be dependent on the stages of amyotrophic lateral sclerosis and the average age of the patient. In the early stages, the patients could handle most communication tools, such as mobile devices, computers, and writing tools. As the course of the disease progresses further, the finer motor skills needed in the use of the mobile phone become a problem. For example, writing a short text message proves to be difficult with progressive paralysis of extremities. The fine motor skills needed for pressing small buttons on the mobile phone to navigate through the menu in order to choose a name in telephone directories and to start a call are diminished, as the motor activity of the hand only allows pressing big buttons. If the linguistic ability is lost, due to progressive paralysis, sometimes a communication system is used. Younger patients use much more modern communication tools than older patients who prefer to use paper-based materials. If paralysis reaches a certain level, patients need assistance from either affiliates or nursing staff.

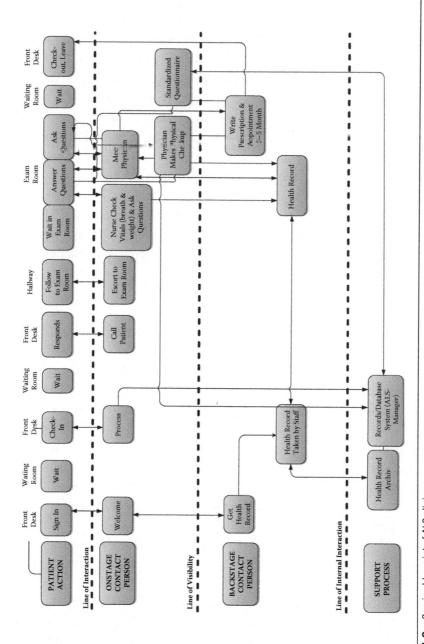

Figure 11.2   Service blueprint of ALS clinic.

Information storage: No arrangement for the regular collection and transmission of nutrition data from a patient's home environment to the medical decision-makers exists. Thus, a progressive loss of weight or dysphagia (swallowing disorder) can remain unnoticed between planned consultations, which means that there are no interventions by physicians, and the treatment processes are not arranged.

There is also at present no acquisition of nutrition data by means of information technology during therapy. Any discrepancy between the actual state of nutrition and the target state in home care is not recorded. This results in further weight loss despite enteral or supplemental nutrition.

Our results confirm the dire demand and need for information and interaction services to improve nutrition tracking for ALS patients. As no solutions for the specific problems of ALS patients have been offered thus far, target processes were elevated and requirements for a nutrition tracking system designed for ALS patients were elicited.

Target State and Requirements Elicitation

The nutrition tracking system needs to aim for increased quality of consultations and perceived quality of life at preferably the same or even reduced costs. This can be obtained by improving the interface problem between the involved actors in nutrition management (Schweiger et al., 2007). By active participation and integration into cure and care processes, better documentation and the resulting improved basis for treatment and nursing can be achieved. Thus, a speedier integration of patients is required. The proactiveness while recording their nutrition status allows patients to gain a deeper understanding of their disease and enables them to cooperate in the treatment and consultation processes.

As ALS patients suffer from motor limitations, they need to be provided with an easy-to-handle nutrition surveillance system that can be used via a mobile phone. Thus, NFC technology has been selected for implementation. NFC is a short-range high-frequency wireless communication technology, based on the frequency of RFID, which enables the exchange of data between devices over about a 10 cm (around 4 in.) distance (ECMA-340, 2004; Want, 2006; Forum, 2007). NFC allows the launching of applications and initiation of actions by simply touching an RIFD tag with an NFC-enabled mobile phone. This intuitive and easy-to-use user interface is especially suited for ALS patients, who are fighting deterioration of fine motor skills with progression of the disease.

The elicited requirements are presented according to the stakeholders:

The patient: The patient plays a decisive and very important role in the management of nutrition. Especially in the earlier stages, the patient can act autonomously and is responsible for his food intake. Later in the course of the disease, he will increasingly be in need of help from his family and care service

Requirements:

- Easy, intuitive, and user-friendly system for the documentation of the daily food intake.
- Possibility to analyze the data on his own.
- Automatic memory signal in case of undernourishment.
- Solution must also be usable in case of progressive paralysis.
- Possible trends in weight development.
- Advice concerning a change in diet.
- Patient control of the access to the collected data by a third person.

Affiliates: The relatives face severe challenges, since they are not only members of the family or friends, but they are often caring relatives. During the course of the disease, they need to deal with massive physical and mental strain.

Requirements:

- A support system for the management of nutrition should help relatives by giving warning signals and concrete options for action, for example, in the form of advice concerning a change in diet.
- Possibly an evaluation feature for the total food input and a function of external monitoring.
- Visibility of trends in the weight development.
- In case of a patient's progressive paralysis, the system must be intuitive so it can be operated by relatives.

Physicians: In general, the physician only sees a patient every three months for the medical examination and for making a diagnosis of the course of the disease. By the time a physician notices a drastic weight loss, the damage can be irreversible. The physician has not been able to recommend a change in diet since there is no information system that documents, monitors, and analyzes the complete nutrition data of patients.

Requirements:

- The system should help the physician anticipate early trends and troubles and therefore prevent a drastic loss in weight.
- The system should allow the observation of the eating habits of patients that is independent of time and place. When necessary, the physician and supportive staff can work together to prevent malnutrition.
- Reception of warnings concerning patients to ensure transfer to appropriate staff.
- Evaluation feature for the total food input and possibility of a long-term analysis for the patient.
- Visibility of trends in weight development.

Nursing staff: In the early stages of ALS, the care services are of lesser importance; however, later in the course of disease, the patients need more intensive care.

Requirements:

- Advice concerning a change in diet and other concrete instructions.
- Easy, intuitive, and user-friendly system for the documentation of the daily food intake if the patient is not able to do this on his own any longer.
- Automatic system warning if the loss in weight is progressive or the patient takes in too little calories.
- Analysis of the quantity and quality of the patients' food intake so that the members of the care service are able to recognize early trends of the patients' weight development. If necessary, the consistency and composition of patients' food can be changed.

To summarize, one of the most important issues is to improve the information logistics between physicians, patients, and nursing staff. An information flow between physicians and nursing staff is especially important to be established, as the sole transmission of information via patients bears certain risks. Further, the position of the physician needs to be considered carefully. As it is a rare and expensive resource in medical processes, real-time surveillance needs to be automated to a high degree so that a cost-efficient care process with improved quality can be realized.

Figure 11.3 illustrates the service blueprint of the target process. This process has been evaluated and reviewed by physicians, clinical staff, and self-help groups. Based on this process, a system design has been developed. In contrast to the current process, it encompasses a higher patient integration enabled by the use of information and communication technologies that enable improved information logistic.

Solution of System Design

The approach presented in this paper is the practical implementation of an NFC-based nutrition management system. This helps the prospective collection of data to detect malnutrition in the early stages.

Patients have access to the platform with mobile devices and computers. They can see the formatted data and can check target-performance comparison if they have taken sufficient calories. If the patients have enabled the analysis of data be given to more people, such as relatives, physicians, or nursing services, who can view the data of the patient, and if necessary, change the nutrition accordingly.

Figure 11.4 shows the system architecture of the NFC-based nutrition management system with a potential connection to a hospital information system (HIS). For selecting the food to be eaten, a touch with the NFC-mobile phone to a picture on the NFC smart poster is sufficient. Once a contact is made, the mobile device gives *audible, haptic*, and *visual feedback* to the patient. The name of the chosen food is spoken (audible), the vibration (haptic) function of the phone is activated, and the chosen food appears (visual) on the display. If the patient wants to send the chosen nutrition to the database, a second touch on the sending tag of the mobile device is enough. Again, the mobile device confirms the transmission of data with audible, haptic, and visual feedback.

An ID is stored on the tag, which is read-out and then compared with data already stored on the mobile phone. Further, in this solution,

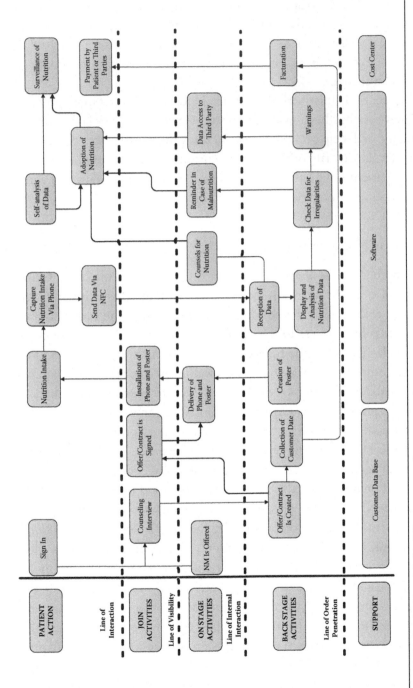

Figure 11.3　Service blueprint of nutrition management.

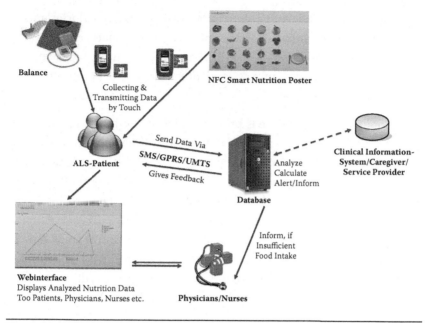

Figure 11.4 Overview of nutrition management architecture.

an NFC-weighing scale is integrated, which sends the measurement data via the mobile device to the database. Depending on the mobile phone contract, the data can be sent to the server via SMS or via a mobile web-based data transaction. Information that is transmitted to the server shows the date, time, the chosen food, and the user ID. This can be done through the mobile device, as well as an identification number that is assigned while installing the software on the device. When transmitting the data of the weighing scale, the same basic data are transmitted, but instead of selecting food data, it will be measurement data transmitted.

The central unit of the solution is a relational database that will store the whole data. The patient and involved persons in the care process have role-based access to the aggregated and analyzed data via the web interface.

The transmitted and aggregated nutritional and vital data are processed and analyzed by rules stored in the systems. If the previously calculated and deposited thresholds of the physician are exceeded, the predefined stages of alert are triggered (see Figure 11.4):

- The system finds small shortfalls in the intake of calories and alerts the patient via SMS, e-mail, or an automatic call.
- After several days of insufficient food intake of patients, selected members, nurses, and physicians will be automatically notified.

The mobile device software is programmed in the Java language. For use on mobile devices, an optimized version of Java 2 Micro Edition (J2ME) is used. The client software stores the multimedia files, such as photos of the food and sound data. Thus, only a small amount of data between the mobile device and the server need to be transferred. The choice of client application, rather than purely server-side architecture, is based on the fact that, especially in rural areas mobile communication networks are insufficiently developed for broadband connection in order to exchange data intensive multimedia files between the server and mobile devices (BMWi, 2009).

Summary and Outlook

Actions and measures that improve nutrition management and the nutritional situation of patients have great economic potential. Although the case of ALS is specific, and there are no reliable figures available, numerous studies do present clinical aftermaths, for example, in the case of the malnutrition of elderly people. These studies document that malnutrition is an independent risk factor that causes not only consequences for the patients' health, but is also responsible for immense additional costs for the public health care system (Stratton et al., 2003, Volkert et al., 2006). Apart from increased nursing and care expenditures, this also encompasses additional hospitalization or medication. According to Stratton et al. (2003), disease-related malnutrition accounts for 16 billion euros of additional costs in Germany alone.

With our nutrition tracking system, we have presented a practical example for the intelligent use of mobile services and information technology to improve the quality of life of patients suffering from chronic diseases. Thus far, we have conducted a requirements elicitation and presented a solution of system design. However, our results present an exploration only and need to be verified in real-world settings. For this purpose, a field study is planned during the next months to evaluate the system with patients and physicians. Beyond this stage, further studies are necessary to evaluate the effects and consequences of an improved nutrition situation for the progression of the disease from a medical perspective.

The example of our nutrition tracking system already shows the potentials and possibilities that technical support systems can offer to patients. During evaluation, it will be essential to test our solution for acceptance by patients, family members, care personnel, physicians, and funding agencies. The obtained results will be integrated into further development processes. In the end, the deployment of technical support systems, such as our nutrition tracking system, can only be successful if the involved stakeholders accept the system, and actively and regularly use it in everyday life. After the questions of usability, the actual benefit to patients, physicians, and nursing staff are subject to close scrutiny.

Additionally, the system needs to be checked for robustness, stability, and scalability. Further, it needs to be evaluated to see whether the technology works in a real world setting and if all processes and organizational issues run and perform as intended or whether they need to be optimized.

Further, an extension of the proposed system to other data and functions is possible. For example, it would make sense to collect data on the current emotional situation or for self-assessment of disease progression. It has already been shown that by completing specific questionnaires, the disease progression in the future can be forecasted (Kaufmann et al., 2005). Another opportunity is offered by the integration of sensor data, especially when the system will be applied in a modified form to other diseases such as multiple sclerosis or adipositas. The extent to which the system can be integrated in existing clinical and medical IT systems also needs to be verified. The required data integration in this context needs to consider new technical and legal developments and conditions in telemedicine and health care. For example, the implementation of new electronic health insurance cards and systems will establish new standards for transmission and storage of data in health care.

To conclude, the use of mobile services and information technology will have a great impact on medical processes and services. In addition to improvements of communication and interaction processes that have a positive influence on quality of patients' lives, cost savings can also be realized. A modified patient–physician relationship, caused by improved patient information and autonomy, offers new possibilities for the design of novel medical services that are only made possible by modern technical support systems. With this system design, our goal is to show an example of the potential of technology-enabled health care processes.

References

BMWi, 2009. *Breitbandstrategie der Bundesregierung.* Berlin: Bundesministerium Für Wirtschaft Und Technologie (Bmwi).

Borasio, G.D. and Pongratz, D.E., 1997. Gedanken zur Aufklärung bei amyotropher Lateralsklerose (ALS). *Der Nervenarzt,* 68, 1004–1007.

Bravo, J., Herv, R., Gallego, R., Casero, G., Vergara, M., Carmona, T., Fuentes, C., Nava, S.W., Chavira, G., and Villarreal, V., 2008. Enabling NFC technology to support activities in an Alzheimer's day center. *Proceedings of the 1st International Conference on Pervasive Technologies Related to Assistive Environments.* Athens, Greece: ACM.

Cleveland, D.W. and Rothstein, J.D., 2001. From Charcot to Lou Gehrig: Deciphering selective motor neuron death in ALS. *Nature Reviews Neuroscience,* 2, 806–819.

Desport, J., Preux, P., Truong, T., Vallat, J.M., Sautereau, D., and Couratier, P., 1999. Nutritional status is a prognostic factor for survival in ALS patients. *Neurology,* 53, 1059–1063.

ECMA-340, 2004. Near Field Communication Interface and Protocol (NFCIP-1).

Fikry, M., Karim, A., and Muhamad, R., 2006. Integration of Near Field Communication (NFC) and bluetooth technology for medical data acquisition system. *Computational Geometry and Artificial Vision,* 147–152.

Forbes, B., Colvilla, S., and Swingler, R.J., 2004. Frequency, timing and outcome of gastrostomy tubes for amyotrophic lateral sclerosis/motor neurone disease: A record linkage study from the Scottish Motor Neurone Disease Register. *Journal of Neurology,* 251, 813–7.

Forum, N., 2007. Near Field Communication in the real world—Part III: Moving to system on chip (SoC) integration. Innovision Research & Technology plc.

Iglesias, R., Parra, J., Cruces, C., and Segura, N.G.D., 2009. Experiencing NFC-based touch for home healthcare. *Proceedings of the 2nd International Conference on Pervasive Technologies Related to Assistive Environments.* Corfu, Greece: ACM, 1–4.

Kaufmann, P., Levy, G., Thompson, J.L.P., Delbene, M.L., Battista, V., Gordon, P.H., Rowland, L.P., Levin, B., and Mitsumoto, H., 2005. The ALSFRSr predicts survival time in an ALS clinic population. *Neurology,* 64, 1, 38–43.

Lahtela, A., Hassinen, M., and Jylha, V., 2008. RFID and NFC in health-care: Safety of hospitals medication care. In C.T.F. Health, ed. *Second International Conference on Pervasive Health 2008*, Tampere, Finland, 241–244.

Leimeister, J.M., Daum, M., and Krcmar, H., 2002. Mobile virtual health-care communities: An approach to community engineering for cancer patients, eds. *Xth European Conference on Information Systems (ECIS)*, Gdansk/Danzig: Wrycza, S., 1626-1637.

Löser, C., Lübbers, H., Mahlke, R., and Lankisch, P.G., 2007. Der ungewollte Gewichtsverlust des alten Menschen. *Dtsch Arztebl*, 104, 3411–3420.

Ludolph, A., 2006. 135th ENMC International Workshop: Nutrition in Amyotrophic Lateral Sclerosis 18–20 of March 2005, Naarden, the Netherlands. *Neuromuscular Disorders*, 16: 530–538.

Meyer, T., 2009. *Wer besucht unsere Sprechstunde?* [online]. http://www.als-charite. de/VM/ALSAmbulanz/Sprechstunde/WerbesuchtunsereSprechstunde/ tabid/257/Default.aspx [Accessed Date 2009].

Morak, J., Hayn, D., Kastner, P., Drobics, M., and Schreier, G., 2009. Near field communication technology as the key for data acquisition in clinical research. *Proceedings of the 2009 First International Workshop on Near Field Communication—Volume 00*. IEEE Computer Society, New York.

Morak, J., Kollmann, A., Hayn, D., Kastner, P., Humer, G., and Schreier, G., 2007. Improving telemonitoring of heart failure patients with NFC technology. *Proceedings of the Fifth IASTED International Conference: Biomedical Engineering*. Innsbruck, Austria: ACTA Press.

Schauder, P., 2006. *Ernährungsmedizin Prävention und Therapie* München; Jena: Elsevier, Urban und Fischer.

Schwabe, G. and Krcmar, H., 1996. Der needs driven approach: Eine meth-ode zur Gestaltung von Telekooperation. In H. Krcmar, H. Lewe, and G. Schwabe, eds. *Herausforderung Telekooperation—Einsatzerfahrungen und Lösungsansätze für ökonomische und ökologische, technische und soziale Fragen unserer Gesellschaft*. Heidelberg: Springer, 69–88.

Schwabe, G. and Krcmar, H., 2000a. Digital material in a political work con-text—The case of Cuparla. In H.R. Hansen, M. Bichler, and H. Mahrer, eds. *Proceedings of the 8th European Conference on Information Systems ECIS 2000*, Vienna.

Schwabe, G. and Krcmar, H., 2000b. Piloting a socio-technical innovation. In: H.R. Hansen, M. Bichler & H. Mahrer, eds. *8th European Conference on Information Systems ECIS 2000*, Vienna, S.132–139.

Schweiger, A., Sunyaev, A., Leimeister, J.M., and Krcmar, H., 2007. Toward seamless healthcare with software agents. *Communications of the Association for Information Systems*, 19: 692–709.

Shostack, L.G., 1982. How to design a service. *European Journal of Marketing*, 16: 49–63.

Shostack, L.G., 1984. Design services that deliver. *Harvard Business Review*, 133–139.

Stratton, R., Green, C., and Elia, M., 2003. *Diseases-Related Malnutrition: An Evidence-Based Approach to Treatment*. Wallingford, UK, 2003.

Volkert, D., Berner, Y., Berry, E., and Al., E., 2006. ESPEN guidelines on enteral nutrition. *Geriatics*. 25: 330–360.

Want, R., 2006. An introduction to RFID technology. *Pervasive Computing*. 6: 25–33.

Yin, R.K., 1989. Research design issues in using the case study method to study management information systems. In J.I. Cash and P.R. Lawrence (eds.) *The Information Systems Research Challenge: Qualitative Research Methods*. Boston, MA: Harvard Press, 1–6.

Zeithaml, V.A., Bitner, M., and Gremler, D.D., 2006. *Services Marketing: Integrating Customer Focus across the Firm*. Boston: McGraw-Hill/Irwin.

12

THE STOLPAN PROJECT

U. BIADER CEIPIDOR, C.M. MEDAGLIA,
A. MORONI, A. VILMOS, AND B. BENYÓ

Contents

Introduction

Over the last 15 years, the growing availability of wireless communication technologies as well as the miniaturization of electronic components inside consumer devices have made it possible to realize a so-called "ubiquitous computing" environment, as foreseen by Mark Weiser at the beginning of the 1990s. In Weiser's article titled "The Computer for the 21st Century" [1], the researcher from XEROX PARC hypothesized a world of interconnected objects by means of information and communication technologies.

In the meantime, another important process related to ICT has been gone through: it deals with the convergence of different communication technologies inside one single device, the mobile phone, nowadays spread among 61% of the worldwide population, according to the International Telecommunication Union (ITU) [2].

The convergence of a number of communication interfaces (Bluetooth, Wi-Fi, W-CDMA, and Near Field Communication [NFC], just to mention some) inside the mobile phone makes this device the most qualified for accessing different types of services in an interconnected environment such as the one predicted by Mark Weiser.

The centrality of the mobile phone in everyone's life is one of the reasons why NFC technology has caught the attention of different industries and research institutions. In fact, NFC can be seen as the integration of Radio Frequency IDentification (RFID) technology inside the mobile phone, potentially allowing more than half of the worldwide population to interact with smart environments via RFID technology.

Although NFC is one of the most promising emerging technologies of the near future, there are still some issues that need to be solved for a mass adoption of NFC technology. While the technical standardization process has been already completed by standardization bodies such as ISO/IEC, ETSI, and also the NFC Forum, application-level standardization still needs to be completed.

The StoLPaN (Store Logistics and Payments with NFC) consortium has worked on creating a single-platform multiapplication environment where many kinds of NFC-based services can coexist and interoperate irrespective of the handset or network specifics, thus creating a homogenous user experience for the customers and a transparent technical environment for the service providers.

Overview of the NFC Environment

When the StoLPaN project started in July 2006, there were great expectations around NFC technology.

The standardization process was almost complete, as initial standards had been in place since 2003: the Near Field Communication Interface and Protocol-1 (NFCIP-1), which specifies "modulation schemes, codings, transfer speeds, and frame format of the RF interface, as well as initialization schemes and conditions required for data collision control during initialization" [3], was prepared by the European association as ECMA-340 and then was published by ETSI (ETSI TS 102 190)

[4] in March 2003, finally approved as an ISO/IEC standard in April 2004 (ISO/IEC 18092:2004). In 2005, the ISO approved also the Near Field Communication Interface and Protocol-2 (NFCIP-2, approved as ISO/IEC 21481:2005 and ECMA 352), which specifies the detection and selection mechanism between reader, card emulation, and peer-to-peer communication modes [5]. Sincere efforts to complete the standardization process were made by standardization bodies.

In 2006, technical standardization seemed to be almost complete, and multiple manufacturers promised several models embedded with an NFC antenna. Analysts from ABI Research predicted that in 2009 half of the mobile phones shipped worldwide would have integrated NFC technology.

In 2007, Nokia launched the first fully integrated NFC mobile phone, the Nokia 6131 NFC, with an embedded Secure Element and GPRS connectivity. Other manufacturers such as Motorola, Samsung, LG, and Sagem proposed their prototypes and, in 2008, Nokia launched the first commercial NFC mobile phone, the Nokia 6212, with embedded SE and UMTS connectivity.

Several pilots were conducted, mainly using Nokia mobile phones, and the results revealed good acceptance of the quickness and convenience of the NFC technology among end users.

Liisa Kannainen, Executive Director of the Mobey Forum, identifies 2005–2007 as the period for technical testing of NFC technology, while 2006–2008 were the years for consumer acceptance testing [6].

Nevertheless, since 2007, several issues related to the NFC ecosystem came to the fore: the secure element's (SE's) position, the related business and operating models, and the value chain for ticketing and payment scenarios. This led to the restructuring of analysts' predictions: in March 2007, ABI Research lowered the number of NFCs shipped worldwide to 30% in 2011.

At the time of this writing (May 2010), the only consumer-available mobile phone equipped with NFC technology is still the Nokia 6212. The delays in shipping NFC mobile phones were due to changes in the handset architecture (SE's position) and in related driving forces (manufacturers versus MNOs). Moreover, the business model is still not clear: who is going to make money and how? Who finances the investments? Who pays for the services? These are still open questions.

Several pilots have been conducted, but they did not resemble commercial operations: in most of the cases, users have been able to test only one or two applications at a time, without the possibility of adding or deleting applications or personalizing them over-the-air.

Regarding NFC use cases, while in 2006 payment and ticketing were the darlings of the industry, the complexity of such scenarios (too many partners involved, too many security issues, etc.) led to the conviction that payment is the last step for an NFC-based implementation, still two or three years away.

The StoLPaN Consortium and Its Contribution to Industry Progress

The StoLPaN consortium has actively worked on overcoming standardization and interoperability issues, mainly dealing with application-level standardization, which has not been adequately taken into account by standardization bodies yet.

The consortium published two white papers concerning the open application and service distribution concept, as well as on the multi-application management concept, endorsed by the NFC Forum and Global Platform.

The first white paper [7] presented a proposal for the postissuance procedures for multiapplication secure elements: it describes the technical model for dynamic card content management of SEs placed in a mobile handset. Card content management deals with the creation and deletion of security domains, application loading, and personalization of smart card applications.

The consortium assumed that, in order to support quick proliferation of NFC services, the industry has to achieve a homogenous, dynamic service environment that would, even after issuance of the cards, allow any service to be loaded onto any SE and managed throughout the application's lifetime.

The card content management process proposed by the consortium includes various roles, enacted by different players, according to the specific situation. This means that, while a single player can enact more than one role, the following functions are necessary to support a dynamic postissuance process:

- *User*: Responsible for initiating the request for postissuance and personalization.
- *Secure Element Issuer*: In charge of controlling the SE, that is, of deciding how the SE's storage capacity is used. He is the owner of the SE's secret keys, which allow him to define who can use the SE, when, and under what conditions. Moreover, he can deploy card content onto the SE.
- *Service Provider (SP)*: Anyone (bank, transport operator, retailer, etc.) who wants to deploy or manage an applet onto the SE. The only requirement to act as an SP is to be compliant with the industry's standard security protocols and SE issuer's specific business conditions.

The StoLPaN consortium has identified also a set of supporting roles that must be considered to provide a fully functional, economic, and convenient service:

- *OTA Provider*: Supplies functions to remotely communicate with the SE, exploiting the real potential of the postissuance and personalization process. From a technological point of view, the communication can rely on different technologies, but the choice will not affect the proposed solution.
- *Trusted Service Manager/Trusted Third Party*: An entity that provides the technology and service support necessary to realize the postissuance procedure in a complex environment, such as the one needed when contactless services are placed on the mobile phone rather than on a smart card. SPs may not be able to operate in such an environment that considers much more roles and relationships between parties compared to a card-based one. That is why a new entity, the TSM, has to be considered to provide a fully functional and convenient service. Even if the involvement of a TSM in the postissuance process has already been considered by the industry, the idea of having more than one TSM participating simultaneously in the process as described by the StoLPaN consortium, represents a completely new approach.
- *Application Issuer*: Guarantees secure interoperability between card and card acceptance device. Sometimes, this role can be enacted by the SP itself (Figure 12.1).

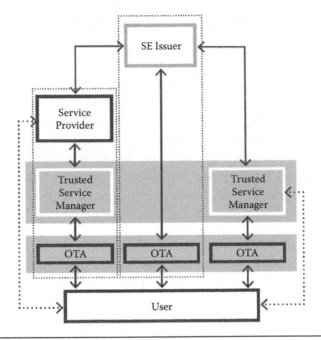

Figure 12.1 Roles in the NFC environment. (From StoLPaN consortium, "Dynamic Management of Multi-Application Secure Elements").

The roles described will probably coexist in several business models, as it is unrealistic to expect one single concept to fit all possible situations. Despite that, the technical process we are going to describe is able to serve and support all the business models that can be placed in the field.

The postissuance process will always start with the user requesting a new service, thus avoiding unsolicited pushed service offerings. Before installing the selected application onto the user's mobile handset, the SP needs to collect details about the NFC device (the end user's device identity), the SE (SE's Card Production Life Cycle information is needed to find the target SE's card issuer and evaluate its security environment), and the SE issuer's contact information, or a pointer to it.

The SP receives this information in the form of a message generated by an application on the user's mobile handset. As an alternative, the information collected can be sent to the TSM instead of the SP, but this operation is transparent to the user as he does not need to know how the service is delivered. If the user's technical environment is compliant with the SP's requirements based on technical, security, and financial considerations, the SP begins the card content management procedure. Otherwise, the SP will inform the user that, for some unidentified reasons, the NFC application cannot be loaded onto the device.

In the case of multiple SEs, the user can express his preference, which the SP has to take into account when choosing where to install the application.

After selecting the target SE, the SP or his TSM partner can trace the SE issuer, according to the information contained in the service-initiating message sent by the user to the SP. This piece of information is the only data element that is currently not available either on the SE or in the mobile phone, and is necessary for starting the automated card content management process.

As the SE issuer is the owner of the keys necessary to access the SE, he is responsible for performing the requested postissuance process, based on the information received by the SP or the TSM. Apart from the generation of security domains and the application installation, the SE issuer also generates specific keys for the SP to ensure exclusive access to the new security domain and related application. To accomplish these tasks, the SE issuer can either use third-party SPs (OTA providers, certification authorities, or TSMs) or perform these tasks himself using his own in-house infrastructure.

Once the SE issuer has loaded required information in the security domain, he sends both a confirmation and the specific keys to the SP (or to the TSM), who can then remotely manage his own application on the user's mobile handset (Figure 12.2).

The presented solution for dynamic NFC application issuance, which represents the first step on the road to NFC interoperability, is entirely described in the StoLPaN's first white paper [7].

The second white paper [8] published by the consortium deals with application-level interoperability, which is the key to NFC's success in our vision, finally enabling the seamless use of NFC services. To define an environment where a dynamic NFC wallet can be created, it is essential to satisfy users, who need a simple way of downloading and removing NFC services; SPs, who need a platform that dynamically accepts their application; and SE owners, who need a platform that helps them to sell SE space in a dynamic manner.

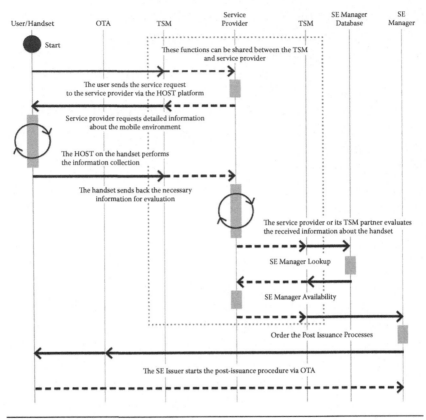

Figure 12.2 Remote postissuance procedure. (From StoLPaN consortium, "Dynamic Management of Multi-Application Secure Elements").

The procedure for application selection is not fully defined in any standard or recommendation, although it is a very complex subject, as the user can have many similar applications (for example, multiple bank accounts with their own application) on the same SE.

In order to provide a seamless user experience of NFC services, the StoLPaN consortium designed and implemented a "Host" platform, which is a wallet application that provides a transparent environment for the simultaneous operation of various NFC-based service applications by neutralizing specifics of the handset design and taking care of resource, security, and communication management.

This application environment concept, together with the application distribution principles explained in the first white paper, create a complete description for an open NFC application environment.

The technical implementation of the Host application will be described in the next section.

StoLPaN Project Results

The StoLPaN project focused on two major research challenges [9]:

- Multiapplication operation in the mobile handset
- Elaboration of the smart retail purchase and payment process

Both these objectives were accomplished, as the StoLPaN consortium has published, and moreover widely circulated, with their underlying theory in the two white papers and has built a proof-of-concept prototype of the Host mobile wallet application. Finally, it has depicted a smart retail scenario, tested in the Libri bookstore and demonstrated to a wide audience at Cartes 2010.

An analysis of the main project results follows.

Host Application

The StoLPaN Host wallet has a modular architecture that relies on general learning coming from the analysis of various NFC use cases (payment, ticketing, access control, loyalty, etc.); as there are common functions between these services, such as the need to store secure data, to give user service information, to allow users to get a new service and so on, the Host is designed to support these functions irrespective of the handset or the operating environment.

Moreover, the Host is based on a component structure, which nowadays is the industry trend (to give an example, consider that in the next generation of mobile Java, MIDP 3.0's architecture is going to be component based). In this way, SPs can dynamically and efficiently integrate NFC applications as components in the Host by means of open APIs, which are available for the programmers.

To go into more detail, the StoLPaN Host interacts with two types of components: the Host Core Components, which are part of the Host itself, and Third Party Software Components (TPSCs), which are not part of the Host. The Core Components provide low-level functions to the Third Party Components, which can be installed, replaced, or deleted without disturbing the Host.

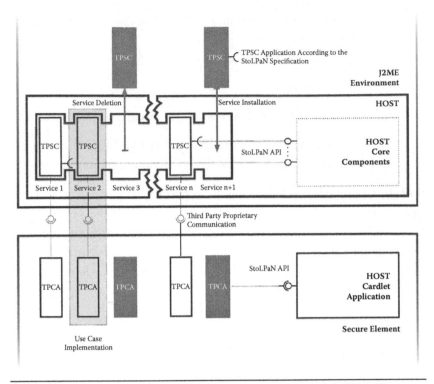

Figure 12.3   The Host architecture (From StoLPaN consortium, "Dynamic NFC Wallet").

The Host Cardlet Application is one of the Host Core Components. It is a smartcard application stored on the handset's SE. The Host Cardlet is in charge of managing security issues in the StoLPaN framework, as it can store keys, certificates, and authentication schemes, moreover providing cryptographic support (Figure 12.3).

Third parties can develop and use their own Cardlets, which store the sensitive application data (user's details, PIN, password, etc.) in a trusted environment provided by the SE. The Third Party Cardlet Applications coexist with the TPSCs, which manage the user interface (UI) and realize the business logic of the third-party-service-specific workflow. They together realize the Third Party Service for the user. They both run on the StoLPaN Host resources, which makes them handset and platform agnostic.

Host and TPSC communicate with the application back office via the StoLPaN Network Protocol (SNP), which sends or receives data over GPRS/3G in J2ME via socket and HTTP connection. The main benefit for SPs is that they can rely on the SNP when building a TPSC, focusing on the application services rather than on low-level communication issues.

Use Cases Developed and Demonstrated

The StoLPaN consortium has developed a prototype proof of concept of the Host application, including sample Third Party Services.

The first services implemented are targeted to travelers: they include a ticketing and an access-locker system. In the ticketing use case, the user can tap the back of his phone to a smart poster for buying one or more tickets for the selected travel zone. The phone reads the tag content and displays product details (zone, price) on the screen, asking the user to fill in the quantity. Then a remote payment is performed, and the ticket is charged into the phone's SE. The user simply taps the back of his phone on the turnstile to open it. Also, the ticket collector can check the validity of the user's ticket by tapping his NFC-equipped device to the user's mobile phone. At any time, the user can check how many tickets are loaded into his mobile phone and eventually buy others.

In the same way, the user can buy electronic keys for opening the baggage locker.

Other use cases implemented are related to the payment scenario. They make use of a mobile wallet loaded into the phone that allows the user to perform both remote peer-to-peer transfers (for example, a mother who transfers some money to her daughter's mobile wallet) and peer-to-peer payments (for paying, for example, the cab fare). Also, purse payments (for example, at vending machines or at a parking meter) are considered, as well as payment transfers and corresponding receiving invoice. Moreover, traditional POS-based payments can be performed with the mobile phone, and the user can also load discount coupons into the SE and use them when performing the payment.

The third scenario demonstrated is related to loyalty programs: the StoLPaN Host supports multiple third party loyalty applications, allowing users to carry only one single device instead of multiple loyalty cards issued by different retailers.

The Smart Shopping Environment

The StoLPaN consortium has developed a complete solution for a smart retail environment, aiming at creating a pleasant shopping experience for the customers, while increasing efficiency for the retail store operators. The StoLPaN solution, described in Reference 10, allows a gradual migration from the already widespread barcode-based solution to the new contactless, NFC-based service.

Technically, the StoLPaN shopping process implements an individual information terminal combined with an individual POS, thus establishing a user-friendly, efficient shopping environment. The core finding is the personalization of the shopping experience, the delivery of personal services to the user's shopping cart, and the removal of check-out and payment counters.

The system is made up of three main components: (1) a smart shopping cart equipped with three pieces of UHF antennas, (2) a Personal Shopping Assistant (PSA) that provides the user interface for customer interaction, and moreover facilitating the remote data exchange with (3) the back office, which is the third component of the StoLPaN solution. It acts as the bridge between the front-end devices and the legacy systems (Figure 12.4).

The shopping cart is provided with a sophisticated antenna system able to read the RFID tags (EPC Class 1 Gen 2) placed on the products inside the cart itself, independently of their position. The antenna signals are then sent to the PSA, which converts them to a protocol manageable by the back office. The communication between the PSA and the back office is established through a WiFi connection. The

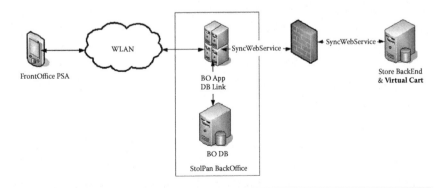

Figure 12.4   Smart shopping environment architecture.

PSA is also provided with an NFC dongle able to read the user's NFC mobile phone or contactless card in order to authenticate it to the system.

The StoLPaN back office has its own business logic, making use of web services to receive any information available in the legacy retail system (product information, loyalty and promotional information, payment data, etc.). To check whether the user has correctly paid for his goods, he needs to push the smart shopping cart through the security gates, which are provided with a UHF antenna system on the top. The StoLPaN back office is in charge of checking whether products leaving the store are identical with the products the user paid for.

To lower implementation costs, the security gates can be replaced by a handheld security terminal carried by the security guards: they can make random spot checks on any of the carts leaving the store.

The aforementioned solution supports the following functionality:

- *Loyalty sign-in*—Customers with a loyalty account (whether it is card or NFC-supported) can sign-in with the PSA. The user can then receive personalized information, such as her loyalty points amount, promotional advertising, or even a shopping list previously filled in at home.
- *Product pricing*—As tagged products are placed inside the shopping cart, the PSA shows the running total on the touch screen. Eventually, it can display a discount price, together with the original price.
- *Product information and location*—Upon placing a product inside the cart, it can be selected for further detailed information. It can include a product description, its pedigree, related products, as well as its position inside the store.
- *E-coupons, discounts*—The shopping cart can read coupons stored on the user's mobile phone or from vouchers, so applying a reduced promotional price.
- *Payment*—Another important function accomplished by the PSA is that it can act as a cashier counter or a point-of-sale terminal. The solution proposed by StoLPaN can support three different payment methods: (1) cash payment, (2) selecting which invoice information is forwarded to a dedicated cashier

desk, and (3) advising the user to proceed to that given counter. In this case, the cashier does not need to scan the products
again, as payment is made on the basis of the invoice generated and forwarded by the StoLPaN back office. The user can
alternatively pay by his bank card, presenting it to the PSA,
which processes the payment information. If a loyalty payment option is selected, the payment is made by using the
preregistered payment instrument of the customer.

A future extension of the service could be the introduction of individual pricing, based on various factors, such as the closeness of the expiration date, the damage of packaging, and so on. It is technologically
possible as RFID tags identify a specific individual product, instead
of generic product category.

Even if the solution described was developed for an RFID-based
environment, it is still possible to adapt it for a barcode-based scenario, allowing a step-by-step migration to the full functional environment. In such a scenario, the PSA is provided with a barcode
reader instead of the RFID antenna, but all the described services
(sign-in, product information retrieval, and coupon redemption) are
still available. Also, at the back office level, the procedures are the
same, so no changes are needed.

Retail Demonstration in Libri The StoLPaN retail environment has
been demonstrated in one of the largest bookstore chains in Hungary:
Libri, with 43 stores and 3.8 million books.

The demo took place in Libri Infopark store, which has an average
of 50 customers per day.

The demonstration included 42 tagged titles with approximately
200 tagged items, replaced when books were purchased. An other
14,000 titles and 40,000 pieces were available with the barcode-based
solution. Ten different coupons with 5%–7% discount were used,
while participating customers with a loyalty card had 10% discount.
To provide continuous help to customers, two hostesses were in the
store full-time to assist the operation for two months, and to fill in
questionnaires with the customers.

Customers could choose between two methods for book selection:
they could use the smart shopping cart with the RFID antenna to

read information on the 42 titles (current bestsellers) equipped with an RFID tag, or use a traditional barcode reader to scan the whole inventory.

Similarly, there were two different modes of operation: customers were lent NFC mobile phones for loyalty, coupon, and payment operation or they could log in to the system with their card for both contactless and barcode-based loyalty and payment procedures.

Participants were not preselected, as anyone coming into the store could test the service. Because of the hi-tech location (Infopark in Budapest), there were probably more tech-savvy customers than the normal average. While a couple of hundred customers tested the solution, more than half of them (111) filled out the evaluation questionnaire. Respondents cover all age groups from under 18 to over 60, and 60% of them were men.

The evaluation methodology was based on both SUMI (Software Usability Measurement Inventory) for evaluating usability and TAM (Technology Acceptance Model) for measuring perceived user acceptance (perceived usefulness and ease of use).

The results showed an overall good acceptance of the solution: there was consent that it would be trendy to have a service like this and people in general would like to use it. Nevertheless, the availability of the technology was strongly questioned, and the majority of the respondents do not expect to see it in less than three years.

The main inhibitor toward adopting a smart retail solution was related to people's habits, as the majority of participants have no experience with using such a technology and do not know what to expect from it. Another barrier that could be quite strong was related to the perceived cost of the solution. Finally, there was no clear risk perception, so more information is necessary in order to make this solution commercially available.

With respect to single function evaluation, payment and product reading were the most preferred ones, while coupon handling with the device was perceived as too complicated and loyalty sign-in was not really appreciated as no plus value was provided. Some additional functions were requested by customers: among them, location information, for example, where a specific book can be found in the store, as well as intelligent product recommendation based on past or present purchases.

A Business Model Proposal

StoLPaN consortium has actively worked on defining a high-level NFC business model outlining both the roles of the leading players within the NFC ecosystem and how funds will be distributed between them. The StoLPaN model aims at being applicable across the whole NFC ecosystem (MNOs, payment institution, transport operators), as well as across EU countries.

The proposed model moves from an accurate review of emerging frameworks for application distribution and from the analysis of the smart transit and payment markets in Germany, France, Hungary, and Italy.

The emerging frameworks analysis has taken into account three important topics of interest within the NFC ecosystem: SE's position, models for sharing it, and application distribution. Regarding the position of the SE, the consortium considered recommendations coming from the North America–based Smart Card Alliance, which promotes SIM card and removable SE, as well as from the European-based GSMA, which recommends use of the SIM card connected to the handset by means of the Single Wire Protocol as SE. With respect to the models for sharing SE's storage, StoLPaN considered the Mobey Forum perspective, detailed in Reference 11, which describes two alternatives for sharing a UICC Secure Element between different third parties: in the hotel model, one single manager, typically the SP, rents the whole space on the SE to third parties, while in the apartment model, the rental is managed by the Trusted Service Manager (TSM).

Regarding application distribution, the following alternatives were considered: a collaborative approach, represented by the Association Européenne Payez Mobile (AEPM), which involves seven banks and four mobile operators in France and started the first-ever NFC collaborative pilot in 2007, sharing TSM functions between multiple organizations; and Global Platform perspective, which employs a single controlling authority that manages key loading and application signing.

The market review considered the payment scenario both in France and in Germany, as well as the ticketing implementations in Italy and Hungary. The mobile payment analysis resulted in the slow adoption of mobile payment solutions, mainly due to cultural barriers: people prefer more traditional payment products such as bank checks, bank transfers, and debit cards. Regarding ticketing services, the main problem is the lack of contactless infrastructure, as there is little commercial desire to migrate to contactless. Support for multiservice propositions (tube, parking, bike sharing, event tickets, payment) is also needed.

What emerged as a key topic is that contactless infrastructures must provide a convenient and trouble-free extension to the existing payment/ticketing products in order to motivate investments and adoption by companies and users.

Conclusion

The StoLPaN consortium has identified four key topics to trace the way forward. They are extensively described in Reference 12.

The first key issue is related to contactless acceptance infrastructure, which is defined as a vital prerequisite for the successful implementation of NFC. With regard to this consideration, it has been noted that in countries where chip cards are already available and widespread, when readers are replaced, the new devices usually have contactless capabilities embedded. Moreover, some banks in the United Kingdom are already issuing payment cards with contactless interfaces.

Another important aspect is to identify appropriate areas for development: the consortium believes that the clearest opportunity for early and successful implementation of NFC is in the area of transit. This is characterized by relatively few stakeholders (compared to payment systems) and good prospects of outside funding (provided by either local or central government), as the perceived social benefits of investing in public transport make this easier to justify than investing in contactless banking infrastructure.

The third topic to consider is related to lessons learnt from existing implementations and trials. In early pilots, the simplest model (a direct agreement between a bank and an MNO) has been used: it works well, but it is limited to a well-defined geographical area. It is hard to see how this type of agreement can be sufficiently scalable and provide the appropriate freedom of choice to customers as NFC becomes more widely accepted. On the other side, the collaborative approach, on which the Payez Mobile initiative is based, represents a more mature ecosystem within a recognized framework, which can be extended to wider areas. As NFC technology becomes more widespread, a standard set of rules could help substantially with cross-border acceptance.

Finally, however, it is clear that a single model will not adequately support all the implementations across entire Europe; a set of generic models can provide a sharable framework from which more specific models can be produced. The single models will take into account local conditions, as well as individual parties' interests. The key consideration in future agreements will be the value each stakeholder can add to the service offered.

The most important factor in the future success of NFC will be the degree to which diverse groups of organizations are able to build on their existing business models to achieve a greater level of cooperation. Once sufficient examples of successful implementations are recognized across Europe, customer demand for greater convenience will become a significant driver. If the demand is sufficient, NFC could be included in monthly mobile packages as a desirable (and chargeable) extra service.

Acknowledgment

The authors would like to thank their partners in the IST-FP6 Project StoLPaN (Store Logistics and Payment with NFC).

References

1. M. Weiser, "The computer for the 21st century," *Scientific American* 265(3): pp. 94–104, 1991.
2. International Telecommunication Union (ITU), "Worldwide Mobile Cellular Subscribers to Reach 4 Billion Mark Late 2008." Available: http://www.itu.int/newsroom/press_releases/2008/29.html
3. ISO/IEC 18092 (ECMA-340). "Information Technology—Telecommunications and Information Exchange between Systems—Near Field Communication—Interface and Protocol (NFCIP-1)." First Edition, 2004-04-01.
4. ETSI TS 102 190 V1.1.1. "Near Field Communication (NFC) IP-1; Interface and Protocol (NFCIP-1) 2003-03." URL: http://www.etsi.org.
5. ISO/IEC 21481. "Information Technology Telecommunications and Information Exchange between Systems—Near Field Communication Interface and Protocol-2 (NFCIP-2)." January 2005.

6. L. Kannainen, "Global Overview of Commercial Implementations and Pilots of NFC Payments during 2009," Smart Card Technology International—globalsmart.com

7. StoLPaN consortium, "Dynamic Management of Multi-Application Secure Elements," Public Whitepaper, available: www.stolpan.com

8. StoLPaN consortium, "Dynamic NFC Wallet," Public Whitepaper, available: www.stolpan.com

9. A. Vilmos, C.M. Medaglia, and A. Moroni, "NFC technology and its application scenarios in a future IoT," *Vision and Challenges for Realising the Internet of Things*, a book published by CERP-IoT in March 2010.

10. StoLPaN consortium, "StoLPaN Smart Shopping," available: www.stolpan.com

11. Mobey Forum, "Best Practice for Mobile Financial Services," Public Whitepaper, available: http://www.mobeyforum.org/

12. StoLPaN consortium, "NFC Application Distribution—Proposed Business Models," Public Deliverable, 2009.

Index

A

Accelerometer, 190

Access control, 30, *See also*
    Authentication;
    Cryptographic protocols;
    Security
    embedded NFC elements, 65
    mobile phone Secure element, 241
    OCBE protocols, 238–240
    tag memory or commands,
        40–41

Actuators, 134, 178

Advanced Encryption Standard
    (AES), 47–48

Advertising applications, 176

AES, 47–48, 238

Affective computing, 280–281

Ambient intelligence (AmI), 154

Amplitude-shift keying (ASK), 28

Amyotrophic lateral sclerosis
    (ALS), 305, 306–308
    course of malnutrition, 310–311
    nutrition management system,
        308–322, *See also* Nutrition
        tracking and management
        system

treatment processes, 311–312
use of materials and tools, 312

Antenna, 27, 28, 32, 34, 35, 60

Anticollision algorithm, 47

Anti-theft systems, 30

ANVIL, 141

Apache Tomcat, 203, 268, 270, 273

Apple, 79

Application design, *See* NFC
    application design

Application programming
    interfaces (APIs), 124

Applications development tools,
    See Mobile applications
    development tools

Association Européene Payez
    Mobile (AEPM), 340, *See
    also* Payez Mobile

Attendance control, 164–166

Audio feedback, 188, 194–195

Authentication, 260–264, *See also*
    Cryptographic protocols;
    Mobile authentication and
    payment service
    challenge-response protocol,
        47–48
    challenges for, 263